KNOWLEDGE-BASED SYSTEM DIAGNOSIS, SUPERVISION, AND CONTROL

T0320502

APPLIED INFORMATION TECHNOLOGY

INDUSTRIAL ARTIFICIAL INTELLIGENCE SYSTEMS
Lucas Pun

KNOWLEDGE-BASED SYSTEM DIAGNOSIS, SUPERVISION, AND CONTROL
Edited by Spyros G. Tzafestas

PARALLEL PROCESSING TECHNIQUES FOR SIMULATION
Edited by M. G. Singh, A. Y. Allidina, and B. K. Daniels

KNOWLEDGE-BASED SYSTEM DIAGNOSIS, SUPERVISION, AND CONTROL

Edited by
Spyros G. Tzafestas
National Technical University
Athens, Greece

PLENUM PRESS • NEW YORK AND LONDON

Library of Congress Cataloging in Publication Data

Knowledge-based system diagnosis, supervision, and control / edited by Spyros G.
Tzafestas.
    p.   cm.—(Applied information technology)
  Includes bibliographies and index.
  ISBN 0-306-43036-3
  1. Expert systems (Computer science) 2. Fault location (Engineering)—Data pro-
cessing. 3. Process control—Data processing. I. Tzafestas, S. G., 1939–   . II.
Series.
QA76.76.E95K57 1989   88-29373
006.3′3—dc19   CIP

© 1989 Plenum Press, New York
A Division of Plenum Publishing Corporation
233 Spring Street, New York, N.Y. 10013

FOREWORD

It is in the area of Systems Diagnosis, Supervision and Control that Knowledge-Based Techniques have had their most significant impact in recent years. In this volume, Spyros Tzafestas has ably put together the current state of the art of the application of Artificial Intelligence concepts to problems of Systems Diagnosis. All the authors in this edited work are distinguished internationally, recognized experts on various aspects of Artificial Intelligence and its applications, and the coverage of the field that they provide is both readable and authoritative. The sixteen chapters break down in a natural way into three broad categories i.e., (a) introduction to the applications of Expert Systems in Engineering, (b) Knowledge-based systems architectures, models and techniques for fault diagnosis, supervision and real time control and finally, (c) applications and case studies in three specific areas, namely: Manufacturing, Chemical Processes and Communications Networks. The final chapter provides a comprehensive survey of the field with an extensive bibliography. The mix of original scientific articles, tutorial and survey papers makes this collection a very timely and valuable addition to the literature in this important field.

MADAN G. SINGH
Professor of Information
Engineering at U.M.I.S.T.

Manchester, England
August 1988

PREFACE

This book provides a set of selected contributions dealing with several issues of system diagnosis, supervision and control through the knowledge-based approach. Since this field is rather new, the major task was to set-up a well-balanced volume involving important representative aspects, techniques and results. I hope that, with the efforts of the eminent contributors who have offered some of their experience, this task was successfully accomplished.

There is a wide divergence of opinion as to which is the best approach to automated on-line fault diagnosis and supervisory control of industrial and other technological systems. The first approach involves the so called algorithmic numerical techniques, such as fault diagnosis using Kalman/Luenberger estimators, diagnosis using cause and consequence analysis, etc. The second more recent approach is to use artificial intelligence and knowledge-based systems techniques which attempt to imitate the reasoning of human fault diagnosers and operators. Knowledge-based techniques provide the means of combining numeric and symbolic models for performing the fault diagnosis and supervisory control tasks. The expert knowledge can be applied to access the plant data and measurements through a distributed control system or from the process control data base.

The importance of reliable automated fault diagnosis, supervision and control of modern technological installations is increasing rapidly as the systems become more complex and they need to operate with minimum malfunctioning or breakdown time. For example, in a chemical industrial plant, product quality is maintained by assuring that process variables fluctuate within permissible ranges. If operating conditions go outside these ranges, the product quality is not acceptable, or more critically some catastrophic event might result. The task of fault diagnosis and process supervisory control is very difficult for the human operator; even well trained operators have difficulties in handling unanticipated events and faults or failures with low probability of occurence. Time is critical in these circumstances, and hesitation, or improper action could lead to disaster. Here is exactly where knowledge-based automated fault diagnosis and supervision can offer a unique aid to operators for detecting, locating and identifying process malfunctions, and for applying high-level supervisory control actions with maximum efficiency.

The book starts with a general introduction to the application of expert systems in engineering. The main body of the book (chapters 2 through 12) presents several knowledge-based system architectures, models and techniques for fault diagnosis, supervision and real time control. Chapters 13 through 15 provide three particular applications namely manufacturing systems, chemical processes and communication networks and, finally, chapter 16 gives a survey of expert systems that are suitable for diagnosis and maintenance of technological systems.

I am deeply indebted to my colleagues for their enthusiastic acceptance to write their chapters and particularly for their effort to provide high-quality up-to-date contributions. As in any volume of the multinational contributed type, there are some variations in the organization and style of the particular chapters. I am convinced that, in spite of this nonuniformity, students, practicing engineers, and research workers will find this book useful in their studies and work.

Athens, June 1988   SPYROS G. TZAFESTAS

CONTENTS

APPLICATIONS OF EXPERT SYSTEMS IN ENGINEERING: AN INTRODUCTION

Sargur N. Srihari

Department of Computer Science
State University of New York at Buffalo
Buffalo, New York 14260

1. INTRODUCTION

A human expert possesses knowledge about a domain, has an understanding of domain problems and has some skill at solving problems. An expert system for a given domain is a computer program that is capable of solving problems at a performance level comparable to that of most human experts in that domain.

Early successes of expert systems were in domains within the fields of medicine (MYCIN diagnoses bacterial infectious diseases) and the sciences (DENDRAL identifies chemical structures). More recently many applications have been found in traditional engineering domains. This is only natural in that engineering, defined broadly, is the specialized knowledge to solve real world problems. Examples of applications of expert systems in engineering are: configuring orders for computers,[1] trouble shooting problems (with diesel locomotives, telephone switching equipment, industrial fan problems, centrifugal pumps, disk drives, utility transformers), process control (chemical plant control optimization), etc.

The objective of this paper is to outline the methodology of first and future generation expert systems in the context of their applicability to engineering and to describe some of the ongoing research on expert systems methodology at the State University of New York at Buffalo.

1.1. Problem Characteristics

The characteristics that distinguish an expert system from other computer programs are that they perform intellectually demanding tasks at expert level performance, that they emphasize domain specific methods of problem solving over general algorithms of computer science, and that they provide explanations for conclusions reached or actions taken.

The kinds of problems that expert systems solve may be divided into those of *analysis* and those of *synthesis*. Analysis problems involve starting with large amounts of data and arriving at conclusions in summary

---

1 XCON configures VAX computer systems at Digital Equipment, similarly
  BEACON is used at Burroughs, CONAD at NCR, and DRAGON at ICL.

form. Examples of analysis problems are interpretation and diagnosis. Interpretation involves classifying or describing data; thus a pattern recognition task such as determining crop categories in a satellite photograph is an expert interpretation problem. Diagnosis is the problem of determining from a set of observables (symptoms, physical finding, results of tests) the causes for the unusual manifestations. Diagnosis in the engineering domain is most often a problem of localization of faults using structural and functional models of the system. Synthesis problems involve beginning with primitive components and arranging or rearranging them to satisfy requirements. Examples of problems of synthesis are configuring a computer, designing a building, designing a mechanical or electrical component, etc.

1.2. Role of Knowledge

Expert problem solving seems to involve search through a judgemental knowledge base specialized to the domain. The knowledge can be either public (published definitions, facts, theories) or private (not in books). Private knowledge is typically in the form of rules of thumb, called *heuristics*, that are learned and refined over years of problem solving experience in that domain. In fact, the central task of building expert systems is that of elucidating and reproducing private knowledge. *Knowledge engineering* is the task of extracting human expert knowledge and organizing an effective representation. The organized knowledge, including heuristics, general models and causal models of behavior is a knowledge base.

2. KNOWLEDGE REPRESENTATION USING PRODUCTION RULES

The method of knowledge representation should have the following characteristics: capture generalization, be understood by people providing it, be easily modifiable, and be useful in a great many situations. Knowledge is represented in first *generation* expert systems in the form of a set of production rules. The basic form of a production rule is:

Rule R_n:

| If | $a_1, a_2, \ldots, a_m$ |
| **Then** | $b_1, b_2, \ldots, b_k$ |

where a_i are predicates (statements that can have true or false values) that are referred to as *antecedents* (also premises, patterns, conditions) and b_i are referred to as *consequents*. The consequents can either be *deductions* or *actions*. A deduction is inferred from facts about a given situation. An example of a deduction rule is (IF(LIGHT ON) Then(CAN SEE)). Deductions are most common in diagnostic reasoning. An action rule changes one situation to another, e.g. (IF(PUMP PRESSURE)>300 Then(SHUT-DOWN PUMP)). Action rules describe expert behavior in terms of available operations.

A rule can be viewed as a conditional statement, and the invocation of rules as a sequence of actions chained by *modus ponens*. According to modus ponens, if A *implies* B is true and A is true, then B is true. For example, given the rule "If x is human then x is mortal" and the fact "Socrates is human", by instantiating x to Socrates and applying modus ponens we have Socrates is mortal. More generally, if A implies B and B implies C, then A implies C — which is referred to as *syllogism*.

When all of the a_i are true, rule R_n is said to be triggered. A rule is selected from the set of triggered rules, or *conflict* set, using a conflict resolution strategy. When the consequents of the selected rule

2

are performed, the rule is said to be *fired*. There are several strategies for selecting the rule for firing from the conflict set. Some of these are:

1. Specificity ordering—arrange rules whose conditions are a super-set of another rule.

2. Rule ordering—arrange rules in priority list; rule appearing earliest has highest priority.

3. Data ordering—arrange data in priority list; rule having highest priority data (condition) has highest priority.

4. Size ordering—rule having longest list of constraining conditions has highest priority.

5. Context limiting—activate (or deactivate) groups of rules at any time; thus there is less likelihood of conflict.

The choice of conflict resolution strategy is ad-hoc. Specificity ordering and context limiting strategies are more often encountered than others.

The different components of a production rule based expert system are shown in Figure 1. The *knowledge base* consists of the set of production rules, *context* is a workspace for the problem constructed by an inference mechanism from the data provided by the user and the knowledge base, and the *inference engine* modifies the context. Apart from the three main modules an expert system should also be provided with a graceful *user interface,* an *explanation facility,* and a *knowledge acquisition* model.

2.1. Forward and Backward Chaining

In the case of synthesis systems the antecedents of the rules are conditions and the consequents are actions. The interpretation or control

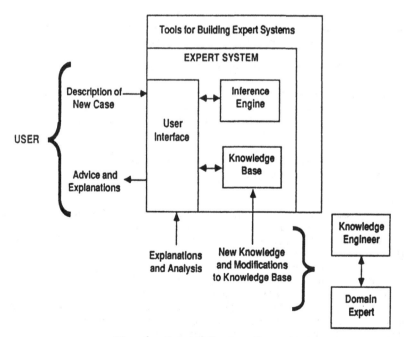

Fig. 1 Expert System Components.

mechanism used by synthesis systems is as follows:

1. Collect rules whose *if* parts are triggered and select a rule using a *conflict resolution* strategy.
2. Do what the rule's *then* part says (fire).

In the case of analysis systems the antecedents of rules are either observed or derived facts and the consequents are new facts that are derived. This mechanism is said to *forward chain* the rules. The control mechanism used by analysis systems can be either *forward* or *backward* chaining. In backward chaining, a particular hypothesis is selected using some discipline. The rules are examined to see if the hypothesis is a consequent. If so, the antecedents of such form the next set of hypotheses. The procedure is continued until some hypothesis is false or all hypotheses are true based on the data.

Forward and backward chaining analysis systems are analogous to bottom-up and top-down control in general computer algorithms, e.g. compilers. Bottom-up analysis systems progressively refine the data to draw conclusions. Top-down analysis systems begin with an expectation of what the data could define and see if the data fits the expectation.

2.2. Certainty Computation

It is often useful or necessary to associate levels of confidence with rules as well as with antecedents and consequents. Thus the inference mechanism needs a method for, say, computing the confidence of the consequent given the confidence values of its antecedents and the confidence of the rule. One approach, based on probabilities, is as follows:

- The certainty of a rule's overall input is the product of the certainties associated with the rule's antecedents.
- The certainty of a rule's output is given by a single valued function having input certainty on one axis and output certainty on another.
- The certainty of a fact supported by several rules is determined by transforming certainties into related measurements, called certainty ratios, then transforming the certainty ratios back into a certainty.

2.3. Production Rule Languages and Environments

In principle, an expert system can be programmed in any programming language such as FORTRAN, C, or LISP. However, as with any complex system, the choice of the tool can influence the feasibility of constructing and/or modifying it. Production rule systems can be efficiently implemented using many different commercial available tools. Prominent among these are: EMYCINS, KAS, ROSIE, OPS5 and PROLOG, among others. Each of these languages/systems provide a built-in inference mechanism.

EMYCIN, a system derived from MYCIN {1}, is particularly suited to diagnosis problems, provides backward chaining, and allows certainties between -1 and +1 to be associated with data and conclusions. KAS was derived from PROSPECTOR, an expert system for geological exploration. It uses likelihood ratios for rule strengths and control can be both forward and backward. OPS5 offers generality in that it is easy to tailor the system to the domain but unlike EMYCIN ans KAS it does not offer sophisticated front ends. OPS5 uses forward chaining exclusively. ROSIE uses English-like syntax but has no sophisticated data base structure. PROLOG is a logic programming language that provides an inference mechanism that is limited to backward chaining. Many production rule systems, e.g. OPS5, are written in LISP (the list processing language most often used in artificial intelligence programming).

2.4. Rule-Based Systems: Pros and Cons

Advantages of a rule-based system are: they enforce a homogeneous representation of knowledge; allow incremental knowledge growth through addition of rules; and allow unplanned but useful interactions — knowledge can be applied when needed and not when the programmer predicts them (by the same token, a disadvantage is that the user can lose control).

Several problems of analysis and synthesis have been successfully handled by *first generation* expert systems which use judgemental knowledge of human experts in the form of a monolithic set of production rules. Among those that have been applied in engineering is XCON {2}, a synthesis expert system for configuring Digital Equipment Corporation's VAX computers and CRIB {3}, an analysis expert system for computer hardware fault diagnosis.

Experience with rule based systems has shown the following drawbacks: knowledge acquisition from domain experts — which is the process of *knowledge engineering* — is time consuming or difficult; all possibilities have to be explicitly enumerated; and they have almost no capability of system generalization.

3. KNOWLEDGE REPRESENTATION USING SEMANTIC NETWORKS AND FRAMES

The keystone to the success of expert systems is the effective representation of domain knowledge. Domain knowledge typically has many forms, including descriptive definitions of domain specific terms (e.g., "power plant", "pump", "flow", "pressure"), descriptions of individual domain objects and their relationships to each other (e.g., "P1 is a pump whose pressure is 230 psi"), and criteria for making decisions (e.g., "if the feedwater pump pressure exceeds 400 psi, then close the pump's input valve").

3.1. Semantic Networks

A semantic network is a method of knowledge representation (see Figure 2). Concepts are represented as nodes (circles) in the network and relations are represented as directed arcs (labeled arrows). A node-and-link net is not necessarily a semantic net, however. To be a semantic network there must be a way of associating meaning with the network. One way of doing this, called *procedural semantics*, is to associate a set of programs that operate on descriptions in the representation.

3.2. Frames

A frame provides a structured representation of an object or a class of objects. For example, one frame might represent an automobile and another a whole class of automobiles. In a sense a frame is a collection of semantic net nodes and slots that together describe s stereotyped object, act, or events (see Figure 3). Constructs are available in a frame language for organizing frames that represent classes into hierarchical taxonomies. In addition, special purpose deduction algorithms exploit the structural characteristics of frames to perform a set of inference that extends the explicitly held set of beliefs to a larger, virtual set of beliefs.

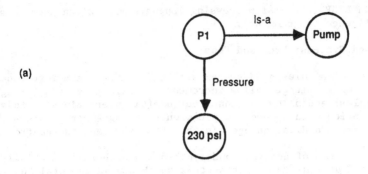

(a)

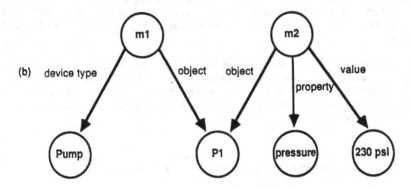

(b)   device type   object   object   value

Fig. 2   A semantic network consists of labeled nodes and arcs
with associated methods of interpreting them: (a) network for "P1 is a pump at pressure 230 psi", (b) equivalent network in SNePS.

3.3. Integrated Environments

Examples of systems that combine the advantages of both frame representation and production rule languages are LOOPS and KEE (Knowledge
Engineering Environment). In these systems, frames provide a rich structural language for describing the objects referred to in the rules.
Frames taxonomies can also be used to partition a system's production
rules.

SNePS is a semantic network processing system that allows relational
knowledge as well as production rules to be represented in the form of a
semantic network. SNePS is particularly appropriate when natural language
interfaces are important.

Future generation expert systems will rely on a problem solving architecture such as the *blackboard,* which is a framework that allows several
knowledge sources (or expert systems) to interact in the solution of a
problem. An environment for such an approach is GBD (Generic Blackboard
Development system).

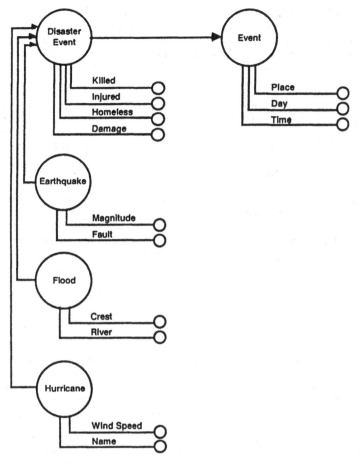

Fig. 3   A network of frames. By inheritance, an earthquake
event has nine slots to be filled.

4. RESEARCH ON EXPERT SYSTEMS METHODOLOGY AT SUNY AT BUFFALO

Research is ongoing in the Department of Computer Science at SUNY at
Buffalo on methodologies for expert systems that solve analysis problems.
In particular we are exploring the design of knowledge-based systems for
image interpretation (recognizing objects in complex environments) and
for diagnosis. In the image interpretation area our work concerns how to
coordinate a community of experts in achieving a common goal {4,5} and
how to develop expert systems for utilizing spatial knowledge in image
interpretation problems {6}. Our image interpretation methods emphasize
the use of a blackboard architecture, frames, and production systems. In
the diagnosis area our work centers on how to represent spatial and other
model knowledge in diagnosis {7,8,9}. In the remainder of this section we
focus only on our research in the area of diagnostic expert systems since
it is a commonly encountered engineering problem.

In the diagnostic application of expert systems first generation systems can also be referred to as *symptom-based,* since the rules consist of
mapping symptoms to intermediate or final conclusions. Expert systems
based on direct mapping of observations to conclusions are said to have

only *shallow* knowledge. Second and future generation diagnostic expert systems that utilize a model of the physical system as an essential part of reasoning may be said to perform on the basis of *deep* knowledge. An important component of deep knowledge consists of design specifications of the device and hence the resulting systems may be said to be *specification-based*.

There are several ways of modeling a physical system. The analytical capability, flexibility, and efficiency of the reasoning process can be expected to depend on the model selected. Two ongoing efforts on modelling and representation for diagnosis are described in the following.

4.1. Behavioral and Structural Models

A physical system can be modeled behaviorally and/or structurally. A behavioral model is a *logical* model and a structural model is a *physical* model. The method of modeling may be different at different levels of abstraction, depending on the task. A logical model is a functionally (causally) oriented abstraction where each logical component has a localized contribution to overall behavior. A physical model is a representation where each component is described in terms of its physical characteristics. In the process of modeling, functional (or logical) structure may be derived from an observation of the physical structure in a bottom-up manner. Two models of the same type have a hierarchical relation if one is an abstraction of the other. The logical and physical models of a physical system often correspond. Thus, for instance, the power train of an automobile consists of two functional (logical) components, the engine which provides power and the transmission which converts power to motion. These, in turn, correspond to the two physical components, the engine assembly and the transmission assembly. However, the two types of models sometime intersect. A digital circuit is functionally modeled by a logic diagram in which logical components such as "and gates" or "or gates" contribute locally to the whole circuit, but several functionally unrelated logical components may be contained in the same physical component, *viz.*, an integrated circuit chip. Our effort if to develop a theory of multi-level device representation within which structural, functional, as well as empirical (symptom) knowledge can be represented.

4.2. Versatile Maintenance

A theory and an implementation of a device representation scheme for versatile maintenance of digital circuits is being developed. Versatility of diagnosis is defined multidimensionally across a range of target devices, faults, maintenance levels, and user interfaces. Central to design model-based fault diagnosis is a device representation scheme with the following attributes: expressive power, ease of building the representation, compactness, generalization ability and expandability, and ability to operate on the representation. Device structure is represented hierarchically to reflect the design model of most devices in the domain. Each object of the device hierarchy has the form of a module. Instead of representing all objects explicitly, an expandable component library is maintained, and objects are instantiated only when needed. The component library consists of descriptions of component *types* used to construct devices at all hierarchical levels. Each component type is represented as an instantiation rule and a structural template. The instantiation rule is used to instantiate an object of the component type as a module with I/O ports and associated functional descriptions. Structural templates describe subparts and wire connections at the next lower hierarchical level of the component type. The implementation of VMES (Versatile Maintenance Expert System) is being done using SNePS.

5. CONCLUSION

There exists a close match between the capabilities of expert systems and the needs of engineering practice. Knowledge acquisition and representation is central to the design of expert systems. Production rules are the preferred method for representing knowledge in first generation expert systems. Second and future generation expert systems will integrate production rules with semantic networks, frames, and other knowledge representation schemes. Our research on expert systems methodology at SUNY at Buffalo is currently focused on problems of analysis such as image interpretation and fault diagnosis. In the diagnosis area the efforts are to develop a theory of multi-level device knowledge representation and an approach to versatile maintenance based on a semantic network representation.

ACKNOWLEDGEMENT

This work was supported by the Air Force Systems Command, Rome Air Development Center, Griffiss Air Force Base, New York 13441-57000, and the Air Force Office of Scientiffic Research, Bolling AFB, DC 20332 under Contract No. F 30602-85-C-0008. This contract supports the Northeast Artificial Intelligence Consortium (NAIC).

REFERENCES

1. E.B. Shortliffe, *MYCIN: Computer-Based Medical Consultations*, Elsevier, New York (1976) {A successful first generation expert system}.

2. J. McDermott, "R1: A rule-based configurer of computer systems", *Artificial Intelligence*, 19 (1) (1982) {Precursor of XCON used at DEC}.

3. R.T. Hartley, "CRIB: Computer fault-finding through knowledge engineering", *Computer*, 17(3) (March, 1984) {An engineering application}.

4. C.H. Wang and S.N. Srihari, "Object recognition in structured and random environments: locating address blocks on mail pieces", *Proceedings of the National Conference on Artificial Intelligence*, (AAAI-86), Philadelphia 1133-1137 (August, 1986) {A vision system using blackboard architecture under development for the U.S. Postal Service}.

5. D. Niyogi and S.N. Srihari, "A rule-based system for document understanding", *Proceedings of the National Conference of Artificial Intelligence* (AAAI-86), Philadelphia 789-793 (August, 1986) {A vision system being developed in collaboration with Xerox Webster Research Center}.

6. R. Kumar and S.N. Srihari, "An expert system for interpreting cranial CT scans", *Proceedings of Expert Systems in Government Symposium*, McLean, (AAAI-86), Philadelphia, August 1986. (AAAI-86), Philadelphia 548-557 (August, 1986, October 1985) {A vision system done in collaboration with SUNY Radiology Department}.

7. Z. Xiang, S.N. Srihari, S.C. Shapiro and J.G. Chutkow, "Analogical and propositional representations of structure in neurological diagnosis", *Proceedings of IEEE Artificial Intelligence Applications Conference*, Denver 127-132 (December, 1984) {Explores spatial knowledge representation issues}.

8. Z. Xiang and S.N. Srihari, "Diagnosis based on empirical and model knowledge", *Sixth International Workshop on Expert Systems*, Avignon, France 835-848 (April, 1986) {Describes a multi-level approach to diagnosis}.

9. M.R. Taie, S.N. Srihari, J. Geller, and S.C. Shapiro, "Device representation using instantiation rules and structural templates", *Proceedings of the Canadian Artificial Intelligence Conference*, Montreal 124-128 (May, 1986) {A theory of device representation under development for VMES}.

Further Reading

10. W.F. Clocksin and C.S. Mellish, *Programming in Prolog*, Springer-Verlag, New York (1981) {An introductory text on PROLOG which is more commonly used as an expert systems development tool in Europe}.

11. R. Fikes and T. Kehler, "The role of frame-based representation in reasoning", *Communications of the ACM*, 28(9) (September, 1985) {The originators of KEE describe its theoretical foundations}.

12. C.L. Forgy, "OPS5 User's Manual", *Technical Report* CMU-CS-81-135, Department of Computer Science, Carnegie-Mellon University, Pittsburgh, PA (1981) {Readable introduction and manual}.

13. P. Nii, "The blackboard model of problem solving", *AI Magazine*, 7: 38-53 (1986) {A tutorial on blackboards to coordinate several knowledge sources}.

14. S.C. Shapiro, "The SNePS semantic network processing system", in N.V. Findler (ed.), *Associative Networks: the Representation and Use of Knowledge by Computers*, Academic Press 179-203 (1979) {Introduction to SNePS by its inventor}.

15. D. Sriram and R. Adey (eds.), *Applications of Artificial Intelligence in Engineering Problems*, Proceedings of the First International Conference, Southampton. England, Springer-Verlag, New York (1986) {Includes papers concerning applications of expert systems in a variety of engineering problems}.

16. P.H. Winston, *Artificial Intelligence (second edition)*, Addison-Wesley (1984) {Excellent introductory text on artificial intelligence that contains a particularly lucid chapter on expert systems methodology}.

A META-KNOWLEDGE ARCHITECTURE FOR PLANNING AND EXPLANATION

IN REPAIR DOMAINS

J. Caviedes*, J. Bourne, A. Brodersen, P. Osborne,
A. Ross, J.D. Schaffer and G. Bengtson+

The Center for Intelligent Systems, Box 1570, Station B
Vanderbilt University, Nashville, TN

* Philips Laboratories, Briarcliff Manor, NY

+ Applied Electronics, Chalmers University
  Gothenburg, Sweden

1. INTRODUCTION

The creation of intelligent systems for use in diagnosis and repair advisory domains has been a popular research topic in the applied artificial intelligence (AI) community during recent years {9,10,16,21}. These systems have been built utilizing a wide variety of architectures. Most early systems focused on the coding of experiential knowledge in the form of if-then style rules {15,18}. Initially, little emphasis was given to the need for incorporation of so-called "deep" knowledge about the physical principles of the systems studied. More recent research has addressed this issue by investigating ways of combining qualitative and quantitative modelling methodologies with inference techniques {17}. However, numerous problems remain to be solved in this domain. For example, little work has been conducted on methods of characterizing control knowledge in diagnosis and repair advisory systems. The research presented in this chapter focuses on this problem. A meta-level architecture that implements a control paradigm that can accommodate both "shallow" and "deep" reasoning mechanisms in a flexible plan-based architecture has been constructed and tested.

In this chapter it will be shown that a meta-level architecture in which the control strategies are declaratively implemented allows a more structured representation of the deep and shallow knowledge levels. Furthermore, with this representation, an explanation system based on a dynamically-generated proof tree naturally turns into a strategy-based explanation system with an inner hierarchy that improves the "friendliness" of the system.

2. THE REPAIR DOMAIN

The building of expert systems to assist in the diagnosis and repair of complex electronic and electromechanical equipment has become quite popular during the last several years. For example, at a recent IJCAI conference, over twenty percent of the papers in the expert systems area addressed this topic. Such popularity is not surprising since, as others have observed {2}, this domain has the required attributes to be an excellent domain for expert systems technology. First, human experts who use many heuristics acquired from experience are available. Second, there is a genuine need for expert systems since a significant fraction of repair and test information about complex devices is not generally available in technical manuals. Finally, given that troubleshooting in the field takes up to 80% of maintenance time {20}, such expert systems may result in significant financial benefits.

Deep versus shallow knowledge is an important issue. Repair-related expert systems are usually built based on shallow knowledge that captures largely experiential knowledge. As these types of systems are refined in order to improve behavior, additional shallow knowledge must be continually added. In contrast, systems built to include understanding of physical laws and principles (i.e., "deep" knowledge) can better adapt to new situations than can systems built using only shallow knowledge. Unfortunately, most repair-related diagnostic systems built today do not understand physical laws and principles. We hypothesize that systems utilizing deep knowledge will evolve rapidly during the coming years.

In this chapter, we have included deep knowledge in the descriptive data structures of the system so this knowledge can be manipulated by the inference engine. The conceptual structure implemented corresponds to a repair-based implementation of a qualitative model {1,19}. Control knowledge, the primary topic in this paper, can be more clearly represented and generalized if it does not get entangled with a non-systematic, mixed representation of domain-specific knowledge and knowledge about general principles.

The solution strategy, the core of the control knowledge in repair domains, can be phrased as follows: A typical technician will approach a repair problem by first ascertaining the major problem, performing one or more tests and pursuing a hypothesis or multiple hypotheses to deeper and deeper levels (i.e., smaller and smaller subsystems) until either the system is repaired or the investigative path no longer seems promising. The hypotheses can be modified so that a best-first strategy can be followed continuously. Compound failures may be detected by use of planning techniques based on sub-goaling, but recursion is mandatory in order to solve non-commutative sub-goals.

With regard to explanation systems, any expert system must have an explanation facility so that (1) the system can explain its actions to the user, and (2) so that the lines of reasoning can be checked during the building and debugging of the knowledge bases. Thus, it is clear that explanations about strategies and plans are necessary to provide a complete and comprehensible explanation. Certainly, the proof tree will assist the user or system implementor, but it will not provide the guiding rationale of a system that can explain strategies and plans. In order to provide explanations of this sort and simplify maintenance of the expert system, our basic premise has been providing an explicit representation of all forms of domain knowledge present in the knowledge base. Such premise is also found in other work on explanation systems, i.e., expert systems generated via an automatic program writer {12}.

In an earlier publication {1}, we described the design of an expert system for use in field service repair domains (Fieldserve-I). This paper presented a conceptual framework for building expert systems in these domains and gave a specific example of a system for assisting field service technicians in board-level troubleshooting and repair. While testing of this system proved its utility, two necessary enhancements became evident. These were: (1) the need for an improved representation of the control mechanism and (2) the ability to provide a more robust explanation to the user. As explained before, these enhancements are interrelated; with the new architecture (called Fieldserve-II) we have achieved both enhancements.

3. META-KNOWLEDGE

The need for explicit representation of control knowledge in expert systems was recognized in the earliest research on expert systems (e.g., {3}). Yet, explicit representation of control knowledge has been almost invariably found in systems that have a performance bottleneck due to the size of the rule base and a weak conflict set resolution strategy (i.e., which rule to fire next). As more complex representations developed, systems have been observed inwhich meta-knowledge was distributed in procedural-style coding throughout the knowledge bases, demons and procedural attachments thus making it difficult to understand the control paradigms incorporated in these systems. It now seems clear that a specific declarative style for representing meta-knowledge is mandatory to avoid confusion and to permit generalization of knowledge at different levels of abstraction.

In the repair problem, as in many others, a simple forward-chaining strategy or a combination of forward and backward chaining encoded procedurally in an agenda is not sufficient to resolve the performance and representational issues described above. The goal of the solution strategy is to re-establish normal operation of the entire system by first specifying the malfunction as precisely as possible, then assigning to it a plausible set of top-level corrective procedures (termed in our work the "check list"). Next, the first of the procedures in the list (usually ordered by subjective likelihood) is executed, and the strategy goes on until the malfunction is corrected or the list is empty. The basic strategy is quite simple, but two characteristics make implementation difficult. First, the check list is likely to change during the consultation because of new evidence or a violation of the current assumptions. Second, the check list typically contains a sequence of sub-goals required to repair the malfunction; hence, non-commutative sub-goals (sub-symptoms) must be solved independently by recursively using the same strategy as for the top-level symptom.

Each of the elements of the check list corresponds to a backward-chained (goal-driven) rule base which, upon correct diagnosis of the problem, suggests a corrective action and verifies proper operation. The knowledge can be shallow or deep, and the quantitative aspects are handled by the inference engine through the appropriate data structures. Meta-rule implementations, although not new {3,6}, usually do not reach the power of the present approach, because either the architecture does not cope well with deep and shallow knowledge or there is not an appropriate protocol for the blackboard management and attention focusing when solving complex repair goals.

The alternative of implementing a control language in a set of procedural predicates by itself or combined with other declarative forms

{11} would not provide the maintainability and accessibility of meta-rules.

4. META-KNOWLEDGE ARCHITECTURE

Figure 1 is an overview of the architecture of Fieldserve-II. In the center of the figure is the multilevel control hierarchy consisting of the following four basic levels: strategy, plan, tactic, and device levels. Different rule bases implement the knowledge at each level as shown by the rule bases below the hierarchy diagram. Several different types of rule bases may exist at each level in this structure - e.g., the repair rule base contains rules which initiate specific repair procedures.

As indicated by the names, the strategy rule base implements a general strategy, the plan rule base is used to formulate a specific plan and the tactic rule base carries out plan actions by running various device level rule bases. A consistency checker evaluates user responses for consistency at the strategy level, and a repair rule base checks to see if a repair has been successfully implemented. Other rule bases that deal with other conceptual repair abstractions can be easily added. For example, a multiple symptom planner can be attached with little difficulty. Intermediate results communicate to the strategy, plan and tactic blackboards. The user interface is realized through messages produced at each level. There is no centralized user interface.

4.1. Control

In the current project, the advantages of expressing control knowledge in the same syntax as the rule-based repair knowledge were explored. There were several design goals for the control structure.

First, the knowledge representation should be specified by rules. Rule format was perceived to enhance knowledge acquisition, application,

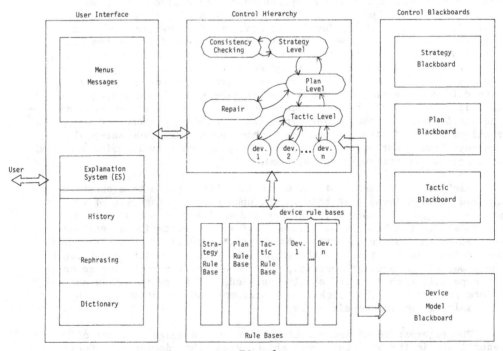

Fig. 1

14

and maintenance. Rule format also provided a uniform architecture which added conceptual clarity to the representation of repair knowledge. Since rules had been clustered according to an overall conceptual structure, the new addition completes it with the control structure which is now more accessible. Second, separate rule bases should be used for different levels of control operations. This feature was specified for the purpose of providing improved explanation. Finally, separation should be maintained between generic and domain-specific control knowledge.

A blackboard with two partitions was implemented. First, the control blackboard stores facts concerned with control of the consultation and is accessible by rule bases at the strategy, plan, tactic, and device levels. This blackboard is separated into three areas: strategy, plan and tactic. Each area can be accessed only by rule bases at the same or adjacent levels. The second blackboard is the "device model", which is used to store all facts acquired about the specific device being repaired. Both structures are implemented as frames.

A consultation is initiated by running the strategy rule base which subsequently invocates the plan rule base. The plan rule base invocates the tactic rule base which controls the running of the device-level rule bases. Control moves between these rule bases depending on the information accumulated during a consultation. Control usually passes many times between the planner, tactician, repairer, and device levels until the problem is resolved or the plan fails. A plan can fail in at least two ways. If the plan stack (a check list) becomes exhausted before the problem is solved, the planner detects this fact and informs the strategy level which exits with an apology. In addition, some of the device-level and repairer-level rule bases contain knowledge of situations that clearly signal the limit of the system's expertise. Such situations can immediately cause an exit through the strategy level with the appropriate apology. In this way, the system can be easily programmed to recognize what it does not know.

4.2. Meta-Knowledge Levels and Rule Description

Each of the levels in the system shown in Figure 1 is described below.

Strategy Level. The strategy level is coded in a single forward-chained rule base which contains the knowledge of the overall device and its externally observable symptoms. For illustration, see in Figure 2 the code and English versions of the strategy rule base. Each consultation begins by activating this rule base. The job of the strategy rule base is to instantiate the blackboard, collect the user's initial observations, check these observations for consistency and call the planner. The strategy rule base also provides the graceful exit from the consultation. The knowledge at this level is device dependent. The organization of the rules is sufficiently general so that changing this rule base for a new device is straightforward.

Consistency Check Rule Base. A rule base for consistency checking is located at the strategy level. Knowledge about legal patterns is available in the consistency check rule base. For example, rules which check to see that the user has not entered conflicting responses are included here In addition, this rule base will infer additional symptoms based on the user input. The complexity of this rule base varies according to the representation language and the inference engine. Some environments have automatic consistency checking performed either based on specifications placed in the data structures and interpreted by the inference engine or through logical inferences made by the inference engine.

```
(strategyRB_rule1   ($type (ifall))
                    ($if ($and
                            ($not (triple_match '(device_model symptoms +)))
                            (triple_match '(control_blackboard strategy
                                            all_fixed no))))
                    ($then (menu_input 'new_device_input)))

(strategyRB_rule2   ($type (ifall))
                    ($if ($and (triple_match '(control_blackboard strategy
                                            all_fixed no))
                            (triple_match '(device_model symptoms +))
                            ($not (triple_match '(control_blackboard
                                            strategy
                                            symptoms_consistent +)))))
                    ($then ($forward 'consistency_checkingRB)
                            (change_slot_to '(control_blackboard strategy
                                            symptoms_consistent yes))))

(strategyRB_rule3   ($type (ifall))
                    ($if ($and (triple_match '(control_blackboard strategy
                                            all_fixed no))
                            (triple_match '(control_blackboard strategy
                                            symptoms_consistent yes))
                            (triple_match '(device_model symptoms +))))
                    ($then (forward 'planRB)))

(strategyRB_rule4   ($type (ifall))
                    ($if (triple_match '(control_blackboard strategy
                                            all_fixed yes)))
                    ($then (change_slot_to '(control_blackboard strategy
                                            all_fixed stop))))

(strategyRB_rule5   ($type (ifall))
                    ($if (triple_match '(control_blackboard strategy
                                            all_fixed cannot)))
                    ($then (change_slot_to '(control_blackboard strategy
                                            all_fixed stop))
                            (msg (n 2) "Better contact Field Engineering!" n)))
```

Fig. 2a. LISP version of strategy-level rule base.

```
RULE  1:  IF   (There are no symptoms chosen)
               (Problem is not fixed)
          THEN (Run the top-level menu input to get symptoms)

RULE  2:  IF   (Problem is not fixed)
               (There are symptoms present)
               (Symptoms have not been checked for consistency)
          THEN (Run rule-base that checks for symptom consistency)

RULE  3:  IF   (Problem is not fixed)
               (There are symptoms present)
               (Symptoms are consistent)
          THEN (Run the PLAN rule base)

RULE  4:  IF   (Problem is fixed)
          THEN (Stop)

RULE  5:  IF   (Problem cannot be fixed)
          THEN (Tell user to contact field engineering)
               (Stop)
```

Fig. 2b. English version of strategy-level rule base.

Plan Level. The plan level rule base contains knowledge for generating and maintaining a plan of action for the diagnosis and repair of the problem at hand. The plan itself is a check list or list of rule base names at the device level, arranged in decreasing order of importance. Initiating, adding to, deleting from and rearranging this list are functions performed by the plan level rules; this check list is naturally placed on the blackboard. The knowledge at this level is device dependent. Due to the hypothetical nature of the sub-goals, this type of meta-planning was preferred to planning techniques like those used by STRIPS {13}. If the inference engine can handle contexts, combining that feature with the planning techniques allows implementation of non-monotonic reasoning techniques as well as symbolic handling of uncertainties.

Tactic Level. The tactic level rule base contains knowledge of how to carry out the plan. The principal action is to pop off the top element of the check list and initiate execution of the corresponding rule base. Each execution results in some status change which is communicated to the tactician. Certain status conditions can be handled by the tactician and some may cause control to be passed back to the planner. The knowledge in the tactician is device independent. When a sub-goal has appeared and is non-commutative in the current context, the sub-goal is solved before continuing by using a method called the subsymptom resolution, analogous to means-ends analysis. Subsymptoms may be detected in rule bases at the device level. A slot is allocated in the tactic entry of the blackboard to signal that a subsymptom has been found. Each subsymptom has its own check list which is processed in the same way as that of the main symptoms. The original plan is then postponed until the subsymptom is fixed, otherwise the consultation fails.

Device Level. Device level rule bases describe specific actions to be taken in diagnosing the problem(s) down to the level of the field replaceable units (FRU's). These rule bases can contain deep or shallow knowledge and will be invoked according to a plan. No general methodology has been utilized to include deep knowledge; however, most sub-goals as well as checks at the end of the check list tend to use deep knowledge more extensively than the rest of the rule bases. The knowledge at the device level is completely device dependent.

The rule bases in this four-level hierarchy communicate with each other primarily through a series of status flags on the control blackboard. In Figure 1 direct invocation of one rule base by another is indicated by solid lines and communication among rule base through the blackboard as indicated by dashed lines. The status flags, as indicated in Figure 3, are "all-fixed" (strategy), "plan ready" (plan), and "current status of item being checked" (tactic). The first entry shown in Figure 3 is the value when the blackboard is initially invoked. Other possible values are indicated in brackets. A design convention is that a rule base associated with one level may only access (read or write) a status slot in adjacent levels.

Repair Rule Base. The repair rule base contains the knowledge of how to modify the plan and/or the device model after a repair action has been successfully completed. A repair has been successfully completed if proper operation was restored as a result of applying the last rule base at the device level. The repair rule base is domain-independent. In our initial implementation there was one repair rule base for each device rule base {7}; however, this was not entirely necessary since device rule bases could verify symptom correction. Yet, modularity is preserved by utilizing the repair rule base, obviating any need for the device-level rule bases to access the control blackboard.

```
Strategy

   All fixed
            no [yes, stop, cannot]

   Symptoms consistent
            [yes, no]

   Number of Symptoms
            n

Plan

   Plan ready
            no [yes, failed, something repaired,
                reorder, stop, replace, remove
                item being checked]
   Consistent
            [names of items to be checked that
             are not yet run]

Tactic

   Current Status of Item being Checked
            not ready [ready, subsymptom found,
                       this item is not the problem,
                       failed, repaired, symptom
                       found, remove]

   Current item being checked
            [name of single item]

   Subsymptom
            [name of subsymptom]
```

Fig. 3 Control blackboard.

Example. Figure 4 shows excerpts from a typical consultation. The
trace facility was "on" in this example to demonstrate the firing of the
control rule bases.

5. HISTORY MECHANISM AND EXPLANATION FACILITY

A history mechanism is implemented that constructs a proof tree.
Each node in the tree corresponds to an important action taken by the
system, such as invoking rule bases, running rules, finding values, etc.
Each node contains specific information about the action represented,
e.g., the result of the action, input parameter value, etc. By examining
this tree, the explanation system can determine why certain rules were
fired, which rules have been tried and where the system is currently
executing. This mechanism is considered to be an addition to the basic
inference engine employed {5,22}.

The implementation of the explanation system (ES) influenced the
architecture that was ultimately employed {8}. Initially, we attempted to
extend earlier work {1} to simply provide a more robust explanation from
the trace of the rule history. The result of this investigation was that
playing back the rule history to the user was unsatisfactory. Indeed, no
information about the rationale for the rule invocation chain was available; hence, the user was often left wondering about the meaning of the
explanation even though the inference chain was presented. Ad hoc procedures which used canned explanations based on the procedurally oriented
and interspersed control knowledge were also tried, but lack of generality
and painstaking maintenance made the approach impractical. These observations led to the formation of the meta-level concept in which strategies
and plans would be explicitly represented in rules rather than in code.

5.1. Object Orientation

To provide a more robust explanation mechanism, the explanation system was designed in an object-oriented fashion, treating entities of the expert system as objects which guide the generation of explanations. For example, rule bases and rules are composed of sets of functional items which are treated as objects by the ES. The nodes of the proof tree are objects and are specially treated depending on the type of object represented by a node. The ES requires that each object guides its own explanation. There is an explanation routine associated with each object which is used during explanation. Thus, objects are responsible for their own explanation. Figure 5 shows the hierarchy of objects. Plan objects are at the highest level that can be explained. Strategies are reserved for system control and thus are not explained.

5.2. Question Restatement

A simple method was implemented to reformat questions asked the user by the expert system. The basic idea was simply to rephrase the question as a preamble to explaining the user's request for explanation. At present, the parser can recognize sentences that start with the following words: are, can, could, has, is, did, does, which, do and was. Depending on the first word of the query, the question restatement system looks for a keyword or subject of the question. The query is restated in a standard format for each of the word types recognized and restated as a preamble for the explanation. The following examples illustrate the method.

> Example 1:
> System   : "Could the board be bad?"
>   {question posed to the user}
>
> User   : "Why?"
>
> System   : "If we can show that the board could be bad"
>   {explanation follows ...}
>
> Example 2:
> System   : "Is it a new unit?"
>
> User   : "Why"
>
> System   : "Why is it necessary to know if it is a new unit?"
>   {explanation follows ...}

The standard restatement of the user question is followed by an interpretation of the "then" parts of the rules used. In cases in which there is a message clause in a rule, the message is simply displayed. When consequent clauses are activated in rules the parser looks for action objects which have standard explanations. For example, in the repair domain the keyword "fixed" is often found. If "fixed" is found on a path on the device model blackboard, the interpretation is that the item has been fixed or not fixed depending on the value concluded. For example, if a rule conclusion (i.e., the consequent part of the rule) is:

> (conclude (device_model fuse fixed) false)

the explanation system would use this keyword to generate the explanation:

> "The fuse is not fixed"

```
-> (consult)
         !1! Applying the strategyRB rule base

         Hi, I'd like to help you repair your equipment

         *** CAUTION ***
         Before beginning to service unit, record all critical data
         stored in volatile memory to prevent inadvertent loss or
         modification.

         *** CAUTION ***
         Always remove power from the unit before you replace any
         board or connector!

              EXPLANATION REQUESTS

         Should you require an explanation when consulting the
         expert system simply enter "why" when prompted.

              PRELIMINARY DATA INPUT

         Before we begin to troubleshoot, take a few
         minutes to verify as many of the following items as
         practical. Enter the ones that are known to be true.

pump_model
    facts
       observed

       1)  Displays appear normal        5)  Control module works properly
       2)  Responds to operating switch  6)  Meets calibration specs
       3)  Energizes power relay         7)  Holds memory
       4)  Computes transaction data     8)  UNIT IS COMPLETELY DEAD

   Please enter choice(s):
     # 1 2

   Now enter the ones that are NOT true

   Please enter choice(s):
     # 3 8

             !2! Applying the planRB rule base

               !3! Applying the tacticRB rule base

                 !4! Applying the C_M_seatingRB rule base

   Was the control module plugged in correctly?

Y/N-> y

                 !4! Leaving the C_M_seatingRB rule base.

                 !4! Applying the try_on_other_unitRB rule base

Try to use the control module on a different unit.
   Does it work correctly on a different unit?

Y/N-> y

The problem must be in the device itself.

                 !4! Leaving the try_on_other_unitRB rule base.

                 !4! Applying the OP_connectorRB rule base

The connection from the Operator Panel to the Logic board
may be loose. Check the connectors and the cable.
   Did you correct the problem by securing the connection?

Y/N-> y

   Did we fix the symptom
   "Control module does not function properly"?

Y/N-> y

Make sure the connection is tightly secured now!

Thanks!
```

Fig. 4

```
                    !4! Leaving the OP_connectorRB rule base.

                  !3! Leaving the tacticRB rule base.

                 !2! Leaving the planRB rule base.

                !1! Leaving the strategyRB rule base.

         End of session

         Do you want to see the complete session log?
           Y/N-> n
         Do you want to store the session log?
           Y/N-> n
         Do you wish to see a summary of the repair operations?
           Y/N->y

                  SUMMARY OF OPERATIONS

         !---!------------------!----------------------------!
         !   !                  !                            !
         !   !  PART            !      OPERATION             !
         !___!_____!_____!
         !   !                  !                            !
         ! 1 ! OP connector     ! secured                    !
         !___!_____!_____!
```

Fig. 4 (continued).

In this example, "device_model fuse fixed" is a path which specifies the location of the conclusion "false".

   Currently the system contains keywords such as: repaired, status, correct, restored, etc. The syntax, however, is not unique, for instance,

(conclude (device_model fuse) fixed)

is the same as

(conclude (device_model fuse fixed) true)

and care must be exercised to insure a consistent representation strategy during knowledge encoding.

   The ES provides a facility for tracing the rule path back through the device, tactic and plan levels, providing explanations by printing the explanation associated with the object at each level in the hierarchy.

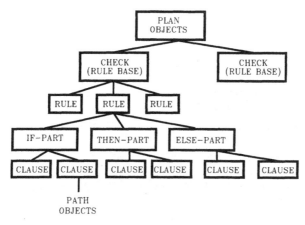

Fig. 5

6. USER INTERFACE AND COMMAND OPTIONS

The explanation system (ES) is entered by the user requesting explanation or assistance by typing WHY at any point during an interactive consultation. For example, Figure 6 shows the system querying the user about the status of a fuse. The user responds "why" and the reason for the question is given. After the initial explanation, the user is again given a chance to invoke any of the available options or to return to the original question. The available options are: WHY, HOW, RULE, PLAN, HELP, EXIT, and SYSTEM. Each is explained below.

WHY Option. When the user initially enters WHY, the current rule is explained. If the user issues WHY as a response to "Option:" the history tree is examined to find the node which caused the previously explained rule to be initiated, and the explanation routine associated with this node is invoked. The user may continue to ask WHY. In this case, higher and higher nodes in the tree will be visited and explanations given .

In order to explain complex rules, the ES breaks up the rule as follows. Suppose a,b,c,d,e, and f are clauses of the If part of a rule structured as

```
(and (not (or a b))
     (or c d)
     (not (and e f)))
```

Also, suppose the rule writer has indicated when constructing the rule that fact c is not to be explained. Given this situation, the above rule clause is simplified to

```
(and (not a)
     (not b)
     d
     (or (not e)
         (not f)))
```

Reconstruction of the rule in this way makes it possible to explain "not" clauses simply as the negation of the clause.

HOW Option. The user enters HOW in response to the option prompt in order to gain an understanding of how a collection of conclusions can be asserted. The If part of the current rule is displayed along with any other information. Figure 7 shows an example.

```
        Is the fuse blown?

    Y/N why

    Explanation requested.

    < 1 > Why is it necessary to know if the fuse if blown?

    If we can show that the fuse is blown, this will help
      us conclude:
      Replace the fuse
      Something has been repaired

    This will help us find out whether "fuse_blown" is the
      problem with the symptom "Missing_voltage_supply."

    Option:
```

Fig. 6

```
Option: how

< 3 > We can conclude that:
  Replace the voltage_regulator on board 1
  voltage_regulator has been replaced
  Missing_voltage_supply has been fixed

by satisfying the following conditions:
  AND Power_supply is not defective
      The fuse is not blown
      120VAC is ok
  OR  The voltage at pin 1 is not 12V
      The voltage at pin 6 is not 12V

otherwise, we can show:
  the voltage_regulator is not the problem

Option:
```

Fig. 7

RULE *Option*. One of the reasons for implementing an explanation
system is to aid the knowledge engineer in debugging the expert system.
To facilitate debugging, the RULE option displays the most recently explained rule in the original syntax that is processed by the inference
engine.

PLAN *Option*. At any point, the user can examine the current plan
that the system is using. Since control dynamically bounces between levels, the plan displayed will be the plan currently active when the
request is made. Figure 8 shows an example of a plan explanation.

HELP *Option*. A help facility is provided for the options available.
Figure 9 displays the screen display for this facility.

EXIT *Option*. Upon entry to the exit option, the explanation session
is ended and control returns to the query being asked when the user originally entered the explanation request.

SYSTEM *Option*. If this option is entered, the user can access all
of system facilities implemented with the inference engine, e.g., rule
and frame editing. This option was included for system maintenance and
is usually employed by knowledge engineers.

Implementation

The system described in this chapter was implemented using GENIE, a
GENeric Inference Engine developed at Vanderbilt University{5}. GENIE is
written in FRANZ LISP {4}, and presently runs on several computers: VAX
11/785, Tektronix 4404, and Apollo. GENIE represents all knowledge in

```
Option: plan

< 4 > We are checking to see whether "Ribbon_cable"
  could be the cause behind the problem "No power to
  board 3" as the next step in our diagnostic plan

With your help, we have already checked Power_supply,
  fuse_blown, and 120VAC, and found that they were not
  the cause of the problem.

The remainder of my plan (if we find that Ribbon_cable
  is not the problem) is to try and see if Wiring,
  or Replace_board3 might be the problem.

Option:
```

Fig. 8

```
Option:  help

Help screen for explanations

why [n] - Explains the "why" reason behind the last screen.
          Optionally explains the "why" reason behind
          screen n.

how [n] - Explains the "how" reason behind the last screen.
          Optionally explains the "how" reason behind
          screen n.

plan    - Gives an overview of the current plan.

help    - Provides help.

rule    - Displays the rule in a GENIE rule format.

exit    - Return to the expert system.

Option:
```

Fig. 9

frames: rules, parameters, blackboards, explanation, and history information. Demons are permitted in frames. System parameters and models of
system devices are stored in frame structures. The use of GENIE in the
field service domain is described in {1}. Further, the results of implementation of the Fieldserve-I advisor are described in {1}.

7. DISCUSSION

Expert systems are frequently criticized for their inflexibility.
Often systems are created that cannot dynamically change their plan of
action. In our research, this rigidity was found to be inadequate for
dealing with repair domains which mandate rapid replanning when moving
from one element to another in the device being repaired. Hence, we implemented the dynamic control strategy using rules described in this chapter. The multilayered architecture facilitates dynamic control by providing all control information in rule form. In our previous implementation, strategies and planning activities were embedded in LISP procedures.
Coding all control information in rules leads to easy maintainability and
understandability of the system; plus, this explicit representation
greatly improves explanations generated from the proof tree. Further, the
encapsulation of specific knowledge about devices in individual rule bases that are controlled from the tactic level provides the benefit of
enhanced understandability of the domain knowledge. Also, device level
rule bases can be easily added, whether they contain deep or shallow
knowledge.

The ultimate objective of our research is twofold. The first is the
generalization of diagnostic and repair problem solving knowledge through
explicit characterization of control knowledge in its two components
(i.e., domain-specific and domain-independent knowledge). The second is
the continuation of basic research on knowledge representation that ultimately should yield more robust problem solvers and simpler knowledge
acquisition techniques.

REFERENCES

1. M. Hofmann, J. Caviedes, J. Bourne, G. Beale and A. Brodersen, Building Expert Systems for Repair Domains, *Expert Systems, The International Journal of Knowledge Engineering*, 3(1):4-12 (January, 1986).

2. K. Dejong, Expert Systems for Diagnosing Complex Systems Failures, *Sigart Newsletter* (July, 1985).

3. R. Davis, Use of Meta-Level Knowledge in the Construction and Maintenance of Large Knowledge Bases, *Stanford AI Memo 283* (July, 1976).

4. J.K. Foderaro, K.L. Sklower and K. Layer, *The FRANZ LISP Manual*, Regents of the University of California (June, 1983).

5. H.S.H. Sandell, GENIE User's Guide and Reference Manual, *Technical Report #84-003*, Electrical and Biomedical Engineering, Vanderbilt University, Nashville, TN (July, 1984).

6. M.P. Georgeff, Procedural Control in Production Systems, *Artificial Intelligence*, 18:175-201 (1982).

7. P.H. Osborne, Meta-Level Control Strategies for Intelligent System Repair, *M.S. Thesis*, Vanderbilt University (1985).

8. A.R. Ross, EASIE: An Explanation System for Expert Systems in GENIE, *M.S. Thesis*, Vanderbilt University (1985).

9. S.C. Shapiro, S.N. Shrihari, M. Taie and J. Geller, Development of an Intelligent Maintenance Assistant, *Sigart Newsletter*, 92:48-49 (April, 1985).

10. R. Davis and H. Shrobe, The Hardware Troubleshooting Group, *Sigart Newsletter*, 93:17-20 (July, 1985).

11. L. Friedman, Controlling Production Firing: The FCL Language, *IJCAI-85*, Los Angeles, CA, pp. 359-366 (August, 1985).

12. R. Neches, W.R. Swartout and J. More, Explainable (and Maintainable) Expert Systems, *IJCAI-85*, Los Angeles, CA, pp. 383-389 (August, 1985).

13. N.J. Nilsson, *Principles of Artificial Intelligence*, Tioga, Palo Alto, CA (1980).

14. P. Fink, Control and Integration of Diverse Knowledge in a Diagnostic Expert System, *IJCAI-85*, Los Angeles, CA, 426-431 (August, 1985).

15. J.S. Bennet and C.R. Hollander, DART: An Expert System for Computer Fault Diagnosis, *IJCAI-81*, 843-845 (August, 1981).

16. R. Davis, et. al., Diagnosis Based on Description of Structure and Behavior, *Artificial Intelligence*, 24(1-3):347-410, 1984.

17. J. De Kleer and B.C. Williams, Reasoning About Multiple Faults, *AAAI-86*, 132-139 (August, 1986).

18. H. Shubin and J.W. Ulrich, IDT: An Intelligent Diagnostic Tool, *AAAI-82*, 290-295 (August, 1982).

19. J. Caviedes and J. Bourne, Knowledge Engineering in Repair Domains: A Characterization of the Task, *6th International Workshop on Expert Systems and their Applications*, Avignon, France, 479-490 (April, 1986).

20. J.G. Wohl, Maintainability Prediction Revisited: Diagnosis Behavior, System Complexity, and Repair Time, *IEEE Trans. Sys., Man, and Cybernetics*, SMC-12(3):241-250 (1982).

21. J.R. Richardson, *Artificial Intelligence in Maintenance*, Noyes Publications, Park Ridge, NJ (1985).

22. G. Bengtson, A History Mechanism for Explanation in GENIE, *Technical Report #85-01*, Department of Electrical and Biomedical Engineering, Vanderbilt University (December, 1984).

3

EXPERT SYSTEMS FOR ENGINEERING DIAGNOSIS: STYLES, REQUIREMENTS FOR TOOLS, AND ADAPTABILITY

Tao Li

Department of Computer Science, The University of Adelaide
G.P.O. Box 498, Adelaide, Australia

1. INTRODUCTION: EXPERT SYSTEMS AND ENGINEERING DIAGNOSIS

Diagnosis seems a natural area for the application of expert systems. A diagnostic problem is often considered a challenge to engineers. They must exercise their intelligence to solve such a complicated puzzle. In solving a diagnostic problem, one must first observe and then try to locate possible failures by proper reasoning. The reasoning can be either empirical (by using accumulated experience) or functional (by using knowledge about system components and organization). Depending on the complexity of systems, their fault diagnosis can be quite complicated and time-consuming. Conventional methods do not seem suitable for sophisticated diagnostic problems. But expert systems can be quite effective in tackling a wide spectrum of diagnostic situations.

As we are aware of, an experienced engineer is capable of diagnosing a fault in much shorter time than an inexperienced engineer. This is because the former has some accumulated knowledge about a certain subject. Frequently, the accumulated knowledge can be formulated as a set of rules. If we collect this set of rules, we can build an expert system. On the other hand, in solving a diagnostic problem engineers also use functional knowledge about a system and its components. When someone encounters a problem in a system for the first time, he or she has no accumulated experience about the system. In this situation, functional knowledge about the structure of the system and the behavior of its components must be utilized to solve the problem. Hence, we can summarize the general knowledge of component behavior as a set of rules and apply then, together with the structural description of a system, for diagnostic purposes. Such a system which employs the knowledge about components and structural description is often referred to as an expert system with functional reasoning.

In this chapter, we discuss the applications of expert systems to engineering diagnostic problems. Section 2 discusses the style of typical diagnostic reasoning. Early diagnostic systems use empirical reasoning, and some new systems use functional reasoning. The empirical reasoning makes the diagnosis process quick and effective, whereas the functional reasoning makes the diagnosis flexible and accurate. A brief discussion on both the *empirical reasoning* and *functional* style of diagnosis will be given in this section as well as some examples. In section 3 we examine

the nature of typical engineering deagnostics. Our investigation will
show that the engineering diagnosis process is typically a mixture of
empirical reasoning and functional reasoning coupled with hierarchical
knowledge. Based on the observations outlined in section 3, we propose
in section 4, certain useful features for the building tools of engineering diagnostic expert systems. These features include a hierarchically
organized knowledge structure, and an inference engine which has explicit
support for making hypotheses. An example is given to show the explicit
support for hypothesizing in one system. The building tools should also
support hierarchical rule organization (but not identical to that of
frame-based systems). Section 5 addresses the issues of flexibility and
adaptability in diagnostic expert systems. Since we would like our diagnostic expert system to be applicable to many tasks, adaptability is very
important. A diagnostic expert system should be portable to other tasks
with minimum effort of reprogramming. Section 6 provides a summary of the
chapter and some future directions for building diagnostic expert systems.

2. THE STYLE OF EXPERT DIAGNOSTIC SYSTEMS

From the literature there seem to be two different styles of diagnostic system design. The first is the so-called *empirical reasoning*,
which has been so named after the second style, i.e. *functional reasoning*,
came to public attention.

In the empirical reasoning systems a diagnosis is obtained through
a set of rules which summarize the expert's experience. This approach
can be very effective in handling routine diagnostic problems. A typical
example of diagnosis with empirical reasoning is the MYCIN system[1,2].
Other medical diagnostic systems and various existing expert systems also
use empirical reasoning.

Example 1

The block diagram of a typical television receiver system is shown in
figure 1. If the syndrome of a malfunction is a horizontal line on the
screen (without a picture), then we might suspect that something is
wrong with the vertical scanning system. Typically this might be a broken
wire leading to the vertical deflection coil, a failed transistor in the
vertical scan output stage, or a short circuited capacitor.

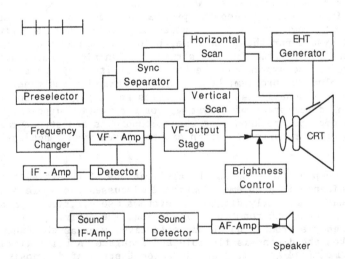

Fig. 1 The block diagram of a television system for diagnosis in Example 1.

Therefore, we can include the following rules for diagnosis.

```
IF picture=one-H-line
THEN V-scan-fault:=true

IF V-scan-fault=true
THEN test(deflect-coil)

IF V-scan-fault=true and
   deflect-coil=normal
THEN test(V-out-trans)

IF V-scan-fault=true and
   deflect-coil=normal and
   V-out-trans=normal
THEN test(V-out-trans)
```

These rules express some experience of the television repairmen. Usually we suspect the parts which have higher failure rates for a particular brand of television. Hence the empirical knowledge is also statistical.

Expert systems similar to MYCIN are referred to as *rule-based* systems. On the other hand, OPS5 type systems are referred to as *production system*. This is because the former can be readily described by its *state space* while the latter can be described by *graph rewriting*. The graph rewriting model(3-4) is not discussed here. Interested readers should see the relevant references. A rule-based expert system can be discribed by the following attributes.

— Let $V=\{v_1, v_2, \ldots, v_n\}$ be a set of variables (parameters) defined for a rule-based expert system. Let $V_I=\{v_{i1}, v_{i2}, \ldots, v_{ik}\}$ be the set of input variables and $V_0=\{v_{o1}, v_{o2}, \ldots, v_{om}\}$ be the set of output variables where V_I, $V_0 \subseteq V$. The values of variables in V_I represent *acquired knowledge*, the values of variables in V_0 represent *solutions*, and the values in $(V-V_I-V_0)$ represent *derived knowledge*.

— Let $D=\{D_1, D_2, \ldots, D_n\}$ be the domain for the variables, where D_i is the set of possible values for variable v_i. (V,D) defines the *state space* of a system.

— $A(t)=\{\alpha_1^t, \alpha_2^t, \ldots, \alpha_n^t\}$, where α_i^t is the value of variable v_i at time t, defines the state of the system at time instance t.

— Let $R=\{R_1, R_2, \ldots, R_p\}$ be the set of rules, where each $R_i:A \to A$ is really a state transition function. Hence the set of rules is actually the transition matrix.

A rule-based expert system can be described by the quintuple (V,D, V_I,V_0,R) which represents the state space and the transition matrix. The behavior of such systems is governed by the *input sequence* and the *transition rules*.

An expert diagnostic system using *functional reasoning* (or, as often referred to, diagnosis from first principles) uses a description of some system including the normal behavior of components, the structural description of the system in terms of its components, and possible abnormal behavior of the components and structural failures. Functional reasoning does not totally rely on expert's experience (statistics, etc.). Instead, it relies on expert's knowledge about the basic principles of a system. A functional reasoning system uses these basic principles for diagnosis. Examples of diagnosis from first principles can be found in the articles of Davis, Reiter and others[5-7].

In several recent articles[7-9], Reiter, DeKleer, and Peng and Reggia gave theoretical formulations of diagnosis from first principles. We will give a brief introduction to diagnonis from first principles. Our definitions are extensions of DeKleer[8], and Peng and Reggia[9].

— A *system description* **SD** is a set of equations which contain information about the relations between the system's components (structural description), and the behavior of the components. So **SD=RC** $\cup$ **BC**, where **RC** is the set of relations between components and **BC** is the set of component descriptions.

— An *observation* is a vector **OBS**=$\{v_1, v_2, \ldots, v_n\}$ where each v_i is the present value of an observation point o_i in the system.

— An *input* is defined as a vector **IN**=$\{u_1, u_2, \ldots, u_m\}$ where each u_i is the present value of excitation at a controllable point I_i in the system. The entire system can be thought of as a state machine.

A *diagnosis*, based on the above definition of a system, is dedined as a conjecture that some minimal set of components are faulty if for the same initial state and the same input sequence IN_1, IN_2, $\ldots$, IN_k, there exists at least one i such that $OBS_i \neq OBS'_i$, where OBS'_i, OBS'_2, $\ldots$, OBS'_k is the output sequence of the faulty system and OBS_1, OBS_2, $\ldots$, OBS_k is the output sequence obtained from the normal system sescription. In a memoryless system, the system state is of no significance. The task of diagnosis is to find such minimal sets of components which could have caused the faulty behavior of the system. Methods of computing diagnosis for memoryless systems are described in depth by Davis, Dekleer, Reiter and others[5-8] and methods for evaluating the plausibility of diagnosis are addressed by Dekleer[7], and Peng and Reggia[9].

Example 2

Consider the circuit shown in figure 2. This circuit is memoryless. If the input is ABCD=1001 and the output is FG=11, we can infer, from the circuit structural description and components behavior, that the output of gate 2

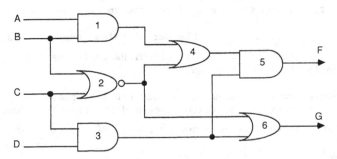

Fig. 2 The digital circuit diagram for diagnosis from first principle in Example 2.

should be a 1. This should also produce a 1 at the output of gate 4 and a 1 at the output of gate 7. The fact that C is 1 indicates a correct output. But the output of gate 5 should be a 0 according to the behavior of an AND gate. Therefore, a fault has occurred. If no information is given for the possible fault occurence in a component, then we can only conjecture that gate 5 is the faulty component. In addition, {gate5} (the set which contains a single component) is a minimal cover which explains the fault. *So {gate5} is a diagnosis.*

On the other hand, if the input is ABCD=1111 and the output is FG=01, we can infer three diagnoses {gate5}, {gate4}, and {gate1}. If the input is ABCD=1111 and the output is FG=00, we can infer that there are four diagnosis {gate5, gate7}, {gate5, gate3}, {gate7, gate1}, and {gate7, gate4}. Note that we assumed that no information is given about the types of faults for each components.

The functional approach to diagnosis is entirely based on the structural description of a system and a complete behavioral description of its components. The behavioral description of components represents the general knowledge about the world and the functional behavior for the system can be derived from such general knowledge and the structural description. In the world of digital circuit testing, there are similar but more specific methods (the cause-effect and effect-cause analysis[10-11] for deriving test patterns.

3. THE NATURE OF ENGINEERING DIAGNOSIS

Engineering troubleshooting and debugging problems seem to conform to a pattern of observation-hypothesizing-measurement and fault resolution. The initial step involves observation. The purpose of this step is to find as many abnormalities as possible to help locate the faults. A set of major parts which may contain some failure components will be suggested according to the initial observation. This narrows down the range of diagnosis. In addition, this step also suggests futher fine measurements for locating the faults. The next step is to perform the measurements and select the faulty component from a set of candidates (this is called "fault resolution"). This pattern can also be observed in medical diagnosis[12-14].

For example, let us consider the working process of a television repairman. He always begins with an observation of basic superficial abnormalities in the indicator lights, the picture display, and the sound channel. When the power is on but the indicator lights are not on and the picture display is dark, he might hypothesize a failure in the power supply circuitry. When the sound channel is working but the picture display is a single horizontal line, he might hypothesize that something is wrong in the horizontal scanning circuitry, typically the output stage. In the hypothesizing step, he uses his experience to approximately locate the source of faults. This requires empirical reasoning. But the empirical knowledge used here is mostly functional which can partially explain the abnormalities.

After he has identified the approximate location of faults, he needs to do some detailed testing and measurements in some major module, for instance, the horizontal scanning circuitry. He would measure the voltages and waveforms at some important points within the horizontal scanning circuitry. An experienced TV repairman would again use his empirical knowledge in this step to test those components which are more likely to fail. The empirical knowledge used in this step is mostly statistical. However, to eventually find the faults he must perform detailed measurements and must use his knowledge about the structure of the module and the behavior of its components to prove that some components are not normal. The final location of faults often requires functional reasoning.

This approach of observation-hypothesis-measurement and fault resolution captures the essence of the engineering diagnostic process. It seems that the majority of engineering diagnostic problems conform to this pattern, and it is thus important to have explicit support for hypohesizing.

The engineering diagnostic process can be described by the same set of parameters as those of functional reasoning systems, with the addition of a set of hierarchically organized hypotheses. Therefore, we use a system description $SD=RC \cup BC$, an observation $OBS=\{\upsilon_1, \upsilon_2,...,\upsilon_n\}$, an input $IN=\{u_1, u_2,...,u_m\}$, and a set of hypotheses $HYP=\{h_1, h_2,...,h_k\}$, where $h_i=\{c_{i1}, c_{i2},...,c_{ip}\}$ and c_{ij} is either a *basic component* or a *subsystem*. In addition, each component c_{ij} is associated with a failure probability p_{ij}. In practice, we would like to arrange the components of a hypothesis in decreasing order of the failure probability. This set of parameters encapsulates the process of engineering diagnosis.

In summary, *empirical reasoning* and *functional reasoning* should be combined in order to effectively accomplish a diagnostic task. Empirical reasoning or functional reasoning alone would not be sufficient. Empirical reasoning alone cannot cover too many possible faults and is not general enough. Functional reasoning alone would require considerable computation and could not avoid the combinatorial explosion, especially for large systems.

4. TOOLS FOR ENGINEERING DIAGNOSTIC SYSTEMS: SOME FEATURES

In the discussion of the previous section we summarized the typical process of engineering diagnosis which often involves a mixture of empirical reasoning and functional reasoning. Another important aspect of engineering diagnosis is the *hierarchical* structure of systems and diagnosis. A system is often organized hierarchically. At each level of the hierarchy, a structural description of a subsystem in terms of its modules and a behavioral description of its modules are given. Hypotheses can be made about these modules and functional analysis can be performed to locate a suspect module.

In our opinion, an expert diagnostic system for real world problems should provide the following features.

1. *Hierarchical organization*. This is not the same as the hierarchical organization in frame-based systems. Property inheritance and other features of frame-based systems are not essential here. All we need is the ability to hierarchically express the structural behavioral knowledge.

2. *Incorporation of empirical reasoning and functional reasoning on each level of the system*. A building tool for expert diagnostic systems should allow for defining hierarchical structures and associating empirical and functional diagnostic rules to each level. The association of rules to levels will greatly reduce the effort of pattern matching and action execution.

In addition to efficiency considerations, we are also concerned with the issue of generating expert behavior in knowledge representation. The appropriate representation will result in better approximation of expert behavior. Some researchers argue that expert knowledge is best encoded in rule, and others argue that structural representation (such as frame-based systems) is a more accurate and robust representation of knowledge. A recent article by Dhar and Pople[15] presented a representation which used a *"model constructor"*. This approach is very flexible. In engineering diagnosis, the expert bahavior is best simulated with a combination of the hierarchical organization and the mixture of *adbuctive* and *deductive* reasoning.

Although we need expert system building tools which support hierarchical organization, and both empirical and functional reasoning, traditional expert system building tools do not have explicit support for engineering diagnosis (in this sense, neither the ordinary rule-based systems nor the frame-based systems can satisfy these requirements). We are facing a lack of proper tools to support engineering diagnosis.

As a first experiment, we have built a tool for implementing engineering diagnostic expert systems. CODAR[16] is the tool we have implemented for this purpose. It partially supports the combination of empirical and functional reasoning. In CODAR, the inference engine operates alternately in two phases, an *abductive hypothesizing phase* and a *deductive inference phase*. However, CODAR does not support hierarchical organization. Therefore the combination of empirical and functional reasoning is realized at only one level. Conceptually, one can build a hierarchically organized system in CODAR. But the implementation would not be as efficient as a system which has direct support for hierarchical organization.

CODAR *is hypothesis-driven,* and its inference process combines both abductive and deductive reasoning. Abduction is one approach to reasoning that is based on hypothesizing. Peirce[17-18] proposed abduction as a *third form of reasoning* in addition to the well-known *deduction* and *induction*. Informally, *abduction* is the process of making hypotheses, *deduction* is the process of verifying hypotheses, and *induction* is the process of establishing general rules from specific facts. Reasoning begins with abduction which proposes hypotheses, followed by deduction which verifies hypotheses, and sometimes induction which establishes general rules from specific facts. The abduction discussed here is not the same as *pure causal reasoning*[19-20], which can be stated as

B and (IF A THEN B) → A

even if it is based on probabilistic selection of causes. In CODAR, abduction is used to realize empirical reasoning.

Abduction must be combined with other forms of reasoning because it only proposes hypotheses. The hypotheses can be verified by deduction. A natural way of combining deductive and abductive reasoning should consist of a *hypothesizing phase,* in which the inference engine makes hypotheses according to some general rules obtained from human experts, and a *verification phase* in which the inference engine employs forward chaining or backward chaining to verify the hypotheses. If the current hypotheses cannot be verified, the inference engine should repeat the cycle again. This two phase cycle may be repeated many times until a hypothesis can be verified or the options are exhausted. This process is implemented in the inference engine of CODAR. The basic structure of CODAR is shown in figure 3.

Some general rules are normally used to guide the hypothesizing phase. These rules are educated guess from human experts and represent empirical reasoning. Furthemore, the rules are not guaranteed to produce verifiable

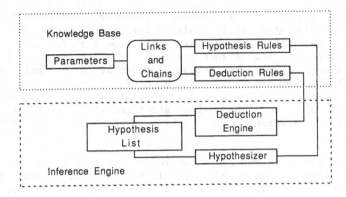

Fig. 3 The organization of the CODAR system. It consists of a knowledge
base and an inference engine. The knowledge base stores variables,
hypothesis rules and deduction rules. The inference engine consists of a hypothesizor and a deduction engine.

hypotheses. CODAR allows the definition of guaranteed hypothesizing rules.

In CODAR both backward chaining and forward chaining are used for
deductive inference. This is to improve the speed of inference and to cut
the search space. Goldberg and Pohl[21] gave a qualitative analysis on the
performance of heuristic search systems.

CODAR is now being extended to include hierarchical organization of
knowledge. The inference algorithm will take into account the hierarchical
structure of a complex system and will allow the association of hypotheses
and rules to every module at each level.

5. ADAPTABILITY OF EXPERT DIAGNOSTIC SYSTEMS

Since we live in a world where technology is rapidly changing, when
we design a new product we must consider the possible changes in the near
future. The design of a product should include things for future expansion.
This is especially important for software products. We would like to
spend the least effort to modify an existing software package for accommo-
dating future modifications. When we build a knoeledge-based system, we
would expect the same flexibility and adaptability of the system. Therefore, it is important to recognize the problem and to incorporate the
adaptability considerations into our knowledge-based systems.

In this section, we briefly discuss the adaptability of various styles
of diagnostic systems. Comparisons between the adaptability of these systems are made and some possible improvements are suggested. To improve
the adaptability of diagnostic expert systems, some basic principles of
software engineering should be observed, among which the *principle of
structural design* is the most important. A diagnostic system designed
structurally and organized hierarchically is much easier to debug and to
modify. This is because the problems under consideration are normally
composed hierarchically. In addition, the building of an expert system can
be considered as a large software project and should obey the principles
of software design.

Obviously, diagnostic systems employing functional reasoning are more
adaptable to various purposes. For example, if we wish to modify an existing system for diagnosis of a similar machine, we only need to update the

data base that contains the structural description of the system, if the set of components are the same for both the old system and the new system. When the two systems have many structural similarities, the modification becomes even simpler. Hence, diagnostic systems employing functional reasoning are very flexible and adaptable.

In a recent article by Dhar and Pople[15], a model of knowledge representation for expert systems was presented. A *"model constructor"* was used to synthesize models that explain and generate expert behavior. A computer program named PLANER was constructed as a result of their research. This approach illustrates the flexibility of expert systems employing functional reasoning. Two kinds of changes were addressed by Dhar and Pople: the first kind occurs within an existing structure, and the second kind occurs in restructuring the structure itself. Therefore, it is necessary to synthesize qualitative models from model parts at each level of abstraction. This is a powerful approach and demonstrates the adaptability of expert systems with functional reasoning.

For diagnostic systems employing functional reasoning, the effort required for modifying the descriptions of basic componet behavior can be readily minimized. This is the main advantage of this approach. However, if the systems is not hierarchically organized, the effort involved in modifying the structural description can be tremendous. Therefore, it is important to have such systems hierarchically organized in the first place. It would be easy if we have an expert system tool that supports hierarchical organization.

Diagnostic systems which employ empirical reasoning are normally designed for specific machines. To modify an existing system can be quite a complex tast. Since the set of parameters for two systems (although similar) can vary drastically. This often requires a restructuring of the rule base. Also, there is no easy way to synthesize expert behavior given the system description. This makes modifications difficult. For example, two brands of some electronic product can have totally different test points and totally different voltages and waveforms at the test points. Hence, diagnostic systems employing empirical reasoning are often not so adaptable.

To improve the adaptability of diagnostic systems employing empirical reasoning, the rule base must be divided properly according to the functionality of the rules, as well as to their structural relationship. In addition, rules of more general forms (usually refer to a subsystem) are more adaptable. Hence it is a good practice to introduce more intermadiate (and hierarchically organized) rules. Even if the expert system tool does not support hierarchical organization, it is always helpful to conceptually divide the system for diagnosis into several subsystem levels and organize the rules accordingly. Not only this helps in future modification of the rule base, but also in the design and debugging of an expert system.

In a system which combines both empirical reasoning and functional reasoning, the adaptability lies somewhere in between that of empirical reasoning and that of functional reasoning. To improve the adaptability of such systems, we also need to follow the principle of hierarchical organization. Our experience of building several example diagnostic systems (including a diagnostic system of traditional Chinese medicine {14} clearly indicates the importance of this principle. For a system which combines both empirical and functional reasoning it is essential that the rules are evenly distributed to every level of the hierarchy. Also, it is important that the division of the hierarchy should make the subsystem as much independent of each other as possible (that is, to make the interface between subsystems as clear as possible). This will reduce the effort of modifying the system if certain subsystems are modified.

In summary, to improve the adaptability of a diagnostic expert system it is important to:

1. Organize the system hierarchically (at least do this conceptually).

2. Make interfaces between subsystems clear and try to avoid side effects in programming.

3. Use a building tool which has explicit support for hierarchical organization if possible.

6. SUMMARY

A discussion of various ways of building diagnostic expert systems is presented in this chapter. Two major styles of engineering diagnosis are considered. Each style has its own advantages and disadvantages.

We argue that a combination of the *functional reasoning* and *empirical reasoning* is more appropriate for engineering diagnosis. The combination of functional reasoning and empirical reasoning produces a highly efficient diagnostic procedure and generates more realistic expert behavior. This combination can be realized through *hypothesizing* and *deductive* inference.

Other issues discussed here include our view on the requirements for design tools of engineering diagnostic expert systems and the adaptability of expert systems. These issues are also important for the design of engineering expert systems and for the applications of expert systems in general.

REFERENCES

1. B. G. Buchanan and E. H. Shortiffe (eds), Rule-Based Expert Systems: the MYCIN Experiments of the Stanford Heuristic Programming Project, *Addison-Wesley*, Reading, Mass. (1984).

2. A. van Melle, System Aids in Constructing Consultation Programs, *UMI Research Press*, Ann Arbor, MI (1981).

3. M. F. M. Tenorio and D. I. Moldovan, Mapping Production Systems into Multiprocessors, *Proc. Int. Conf. Parallel Processing*, pp. 56–62 (1985).

4. H. Erhig, Introduction of Algebraic Theory of Graph Grammars, *LNCS* Vol. 73, *Springer-Verlag* (1978).

5. B. C. Williams, Qualitative Analysis of MOS Circuits, *Artificial Intelligence*, 24, pp. 281–346 (1984).

6. R. Davis, Diagnostic Reasoning Based on Structure and Behavior, *Artificial Intelligence* 24, pp. 347–410 (1984).

7. R. Reiter, A Theory of Diagnosis From First Principle, *Artificial Intelligence*, 32, pp. 57–95 (1987).

8. J. de Kleer and B. C. Williams, Diagnosing Multiple Faults, *Artificial Intelligence*, 32, pp. 97–130 (1987).

9. Y. Peng and J. A. Reggia, Plausibility of Diagnostic Hypotheses: the Nature of Simplicity, *Proc. AAAI-86*, Philadelphia, P.A., pp. 140-145 (1986).

10. M. Abramovici and M. A. Breuer, Multiple Fault Diagnosis in Combinational Circuits based on An Effect-Cause Analysis, *IEEE Trans. Comput.* Vol. C-29, No. 6, pp. 451-460 (June 1980).

11. D. C. Bossen and S. J. Hong, Cause-Effect Analysis for Multiple Fault Detection in Combinational Networks, *IEEE Trans. Comput.*, Vol. C-20, No. 11, pp. 1252 (Nov. 1971).

12. H. Pople, Heuristic methods for imposing structure on ill-structured problems: The structuring of medical diagnosis, In P. Szolovitz, ed., *Artificial Intelligence in medicine.* Boulder Colo: Westview Press, pp. 119-185 (1981).

13. J. A. Reggia, D. S. Nau and P. Y. Wang, Diagnostic Expert Systems based on A Set Covering Model, *Int. Journal Man-Machine Studies*, 19, pp. 437-460 (1983).

14. T. Li and L. Y. Fang, Twoway diagnosis in traditional Chinese medicine, *Proc. 3rd Internat. Conf. Applications of Artificial Intelligence*, Orlando FL, pp. 937-945 (April 1986).

15. V. Dhar and H. E. Pople, Rule-based versus Structure-based Models for Explaining and Generating Expert Behavior, *Commun. ACM*, Vol. 30, No. 6, pp. 542-555 (June 1987).

16. T. Li, "CODAR: A General Purpose Design Tool for Rule-based Engineering Expert Systems", *Journal of Artificial Intelligence in Engrg.*, (1987).

17. W. Burks, Peirce's Theory of Adbuction, *Philosophy of Science*, 13, pp. 301-306 (1946).

18. W. Burks, Chance, Cause, Reason, *University of Chicago Press*, Chicago, IL, (1977).

19. E. Charniak and D. McDermott, Introduction to Artificial Intelligence, Chapter 8, *Addison-Wesley*, Reading, Mass. (1985).

20. J. A. Reggia, Abductive Inference, Proc. Expert Systems in Government Symp., McLean Virginia: *IEEE Computer Soci. Press*, pp. 484-487 (1985).

21. A. Goldberg and I. Pohl, Is Complexity Theory of Use to AI?, *Artificial and Human Intelligence* (edited review papers of Internat. NATO Symp. Artificial and Human Intelligence Lyon France), edited by A. Elithorn and R. Banerji, *North-Holland*, pp. 43-56 (1984).

4

THE USE OF MULTI-DIMENSIONAL FRAMES IN FAULT DIAGNOSIS KNOWLEDGE BASE

Igor Gazdik

Elinsborgsbacken 23, S-163 64 SPANGA, Sweden

INTRODUCTION

Ever since man, in his capacity of homo faber, produced the first tools, weapons and other implements, systems and components reliability has been a factor of paramount importance for the survival of mankind. To assure reliable functioning of man-made artifacts, extend their life and maximize the return on investment into them, reliability assurance (in particular timely and proper maintenance) is necessary. At the heart of maintenance activities lie diagnosing, elimination and prevention of functional faults in engineering systems and their constituent parts.

Fault diagnosis, elimination and prevention are strictly man-centered problem-solving activities. They are interleaved with man's dual role in the production process: man is the producer of material goods, but he also is the consumer of the results of his own production. In his latter role man sets requirements on, and standards for, his activity in his former role. Reliability assurance is a bridge between these two roles. Thus, the success of man's fault diagnosis and prevention activities largely depends on his motivation and ability to assess the problem situation he is to handle, as well as on his skills and imagination in mastering that problem situation. Added to this is the fact that fault diagnosis and prevention work cannot be controled and supervised as closely as production work, because the worker needs a considerable autonomy over his day-to-day actions and decisions. Consequently, a successful fault diagnosis is affected by two factors. One is the human factor represented by a tradeforce that is both diligent and committed, i.e. properly motivated, trained and stimulated. The other is the machine factor; more reliable and more user-friendly machines facilitate fault diagnosis and make it simpler. Both aspects corroborate each other, and both have been researched extensively. But how fruitful, from the fault diagnosis point of view, are the results of these researches?

Several works (1-5) present the results of motivation studies. The most all-round techniques for training operators in fault diagnosis, and sharpening their perception of problem situations early and correctly, are described in (4, 6-8). Stimulation of workers who are sufficiently gifted and properly motivated to seek problems, but deficient in advanced engineering knowledge needed for solving technical problems, is treated in (9). Various financial incentives and organizational measures enhancing the human effort are an important and widely used complement to the theoretical research.

The machine aspect has been tackled in many ways, too varied to be expounded here. They depend on the progress of human knowledge in such fields as materials science, systems theory, environmental and operating conditions, or ergonomics. This progress affects various design improvements, device modularization and the use of field replaceable units. The aim is to reduce the number of repair actions and simplify their implementation in a given system.

In spite of a full-scale application of these achievements of psychological and engineering research, the prediction of unreliability and disturbances in engineering systems, as well as assessment of the length of life of system components, still remain difficult tasks. It is because fault diagnosis and other reliability assurance activities are obscured by a haze of uncertainty - an uncertainty stemming from the man-machine interaction, man's moodiness, fatigue, ignorance or unjustified boldness, vagueness of the human vernacular, incompleteness of the problem assignment, insufficiency of the theoretic techniques available, as well as various social and political considerations. Indeed, man's ability to control these uncertainties affects proportionately his efficiency in fault diagnosis and prevention.

1. EXPERT SYSTEMS

1.1. Expert systems in Failure Diagnosis

A natural approach to mitigating the impact of uncertainty in fault diagnosis is to render fault diagnosis "machine friendly" by making it less dependent on one person's capabilities or momentary state of mind. This has become possible in recent years owing to the progress made in artificial intelligence (AI) research. One of its outgrowths are expert systems - a special kind of software that uses dedicated rules to process concentrated knowledge of many experts, and places it at the user's disposal (10-32). Expert systems (ES) in troubleshooting are expected to speed fault diagnosis up, increase its thoroughness and accuracy, make it more consistent and more readily available to the non-expert user. Their deployment is enhanced by such factors as the high cost of downtime for the equipment being diagnosed, scarcity of human experts, and an availability of relatively well-defined problems

How do expert systems differ from other computer applications? While conventional algorithmic programming, with its steps, tests and branches, follows predefined options, an expert system works from a collection of IF-THEN rules. A part of the system, called the inference engine, recognizes when a particular rule can be applied. As soon as the inference engine finds that the present situation matches the IF-part of the rule, that rule can be activated. It implies that the action embodied in the THEN part of the rule will be taken. The reasoning logic these production rule systems deploy is generally of two kinds: forward chaining proceeds from facts and rules to conclusions; backward chaining starts with tentative hypotheses and seeks supporting facts.

A few years ago expert systems emerged from laboratories and started assuming a position on the factory floor. This immediately introduced new

diversifications in their applicability, taxonomy and hope of their performance. It is therefore natural to ask, given a certain application,
whether one system is as good as any other and, particularly, whether there
are any special requirements placed on expert systems for fault diagnosis.

Not so long ago, these question would have seemed irrelevant, because
in the light of AI research, expert systems are supposed to mimic "human
reasoning" in processing the human expertise they contain. Using a chain
of IF-THEN rules they are expected to come up with a sensible answer to any
question within their field of expertise. Nowadays, those questions are
relevant because, in practical applications, the performance of an ES is
judged simply on the basis of its output, i.e. the system's ability to give
correct answers to a variety of assignments. To give correct, high quality
answers, the system must essentially be able to extract crisp, credible and
dependable knowledge from the contents of the initially incomplete, vague
or ambiguous rules. To achieve this goal, an expert system, including its
underlying theoretical concepts, has necessarily to be individualized with
respect to the unique requirements of a specific fault diagnosis problem,
which cannot be expressed in general terms in the language of AI. Therefore,
in real applications, one ES can be better than the others because it is
more in line with reality.   The following comments will illuminate the
reasons why it is so.

In the theoretical expert system, the connection between faults and
their impact is based on empirical evidence rather than an underlying model
of how the equipment functions. The fault diagnosis software may thus make
obvious mistakes when faced with unfamiliar problems, incomplete evidence,
or multiple failures/influences. If the equipment to be diagnosed is changed
even slightly, the new behavior must be incorporated into new rules in much
the same way the diagnosis knowledge was originally built up. Good troubleshooting calls for more than just the ability to diagnose faults when
given complete information about an equipment's condition; human experts
usually start with incomplete data about a malfunction. The expert system,
faced with uncertain information, may find it difficult to decide which
tests will furnish the most information about unreliable components of the
equipment analyzed. A useful test must pinpoint those candidates that are
likely to fail often. To design an expert system for this purpose is not
an easy task.

One failure that defies automated analysis, and is hard for even the
most sophisticated systems to crack, is the intermittent fault. Because it
does not always occur, it requires reasoning about equipment behavior over
periods of time, something that knowledge based systems can hardly do in
less than a crude fashion.

Time-related behavior is another problem whose full understanding
still eludes researchers in AI. For example, faults may be triggered or
revealed only by a particular sequence of events or a particular configuration of influences acting on the equipment. Dealing with such failures
requires techniques for representing both the order of events, and the
length of time between events (or their duration).

The very inclusion of what is believed to be human-like reasoning
capability (modeled by the postulates of logic) comes under fire. It often
slows program execution and complicates system development. More and more
scientists voice their criticism of using logic as a means of understanding
human thinking. Such criticism is well justified when considering the
continuity of human thinking and the broad spectrum of engineering problems
to which it is applied. Relatively few of these problems can be solved by

using simple logical formulas and empirical facts. To solve advanced engineering problems, an extensive use of theoretical knowledge and mathematical methods is necessary.

Experts for certain assignments may not be available. This is particularly true in situations when the expert system is used as an intelligent interface between the user and the databases used, rather than the traditional stand-alone expert system. One reason for the scarcity or unavailability of experts may be that the ES applications are to support functions that cannot be performed manually. For instance, data may be produced in a rapid succession by various sensor arrays at a rate which requires speeds people cannot handle. Another reason is that the ES may be part of the development of a new system with the consequence that experts with experience in using the basic equipment configuration cannot exist. In many situations it may be difficult to assess who is an expert, resp. to choose the right candidate for a job. Therefore, common evaluation procedures comparing human and machine reasoning are no longer valid.

But, the advantages expert systems offer are too attractive to be sacrificed because of terminological limitations of a scientific discipline. Instead, the scope of expert systems has to be widened by including into them the more extreme forms of expertise. One is the so called heuristic search, i.e. a sequence of educated guesses and hunches, with the confidence in the validity of the guesses increasing at each step. The other is problem solving based on fundamental knowledge.

Reasoning based on fundamental knowledge uses mathematical theorems and axioms - a common approach to problem solving by human experts. Fundamental knowledge ordinarily does not find direct use in expert systems but can be helpful in creative solving of unusual problems. In section 1.3. it will become evident that in special reliability projects this approach is justified.

The design of a usable high quality ES starts with the analysis of the requirements a specific fault diagnosis assignment places on the system in terms of accuracy of the analyses the ES ought to perform. That accuracy can best be achieved by a suitable blend of heuristics and fundamental knowledge which both are part of human expertise. Then it is necessary to decide where in the expert system program will the heuristics, resp. fundamental knowledge, be applied. Ancillary issues, such as the extent of conversation the system can carry out with its user are, in many engineering assignments, often of lesser importance. In fact, in many important applications the conversation is a burden often replaced by graphical means of communication (33).

1.2. RELSHELL - A Knowledge Base for Fault Diagnosis

A useful new tool, largely independent of terminological ambiguities and attempts to stretch the meaning of the word "expert", is the expert system shell. It is an expert system the knowledge base of which is empty. The user can buy a ready made shell (or develop it) and then fill its knowledge base according to the exigencies of the problem situation the shell is to handle. The knowledge base can thus be made to satisfy the requirements of the assignment to be handled without being dependent on the knowledge of an external expert.

An expert system shell for fault diagnosis, called the RELSHELL, was introduced in (34). Its purpose is to identify the origin of various functional faults eventually leading to equipment malfunction. Most of the crisp, credible RELSHELL results follow from some fuzzy-arithmetic or fuzzy-logical operations (35-43). In this paper, the analyzing power of the RELSHELL is extended. But, first a brief recapitulation of its structure.

1.2.1. Problem domain

The RELSHELL was designed to handle degrading multi-parametric equipment (MPE) characterized by the following features:

a. More than one variable is needed to characterize the state of the equipment;

b. The initial state of the equipment is defined by some configuration, adjustments, and calibration of equipment components;

c. Complex linkages and interactions exist between the components of the equipment;

d. Because of the action of some internal and external influences (assumed to be observable), the equipment degrades gradually. The degradation manifests itself as a set of observable and measurable symptoms;

e. The state of equipment failure cannot be defined uniquely (it is not binary). The symptoms of failure are multiple functional faults and difficulties caused by components drifting gradually out of calibration;

f. To assure reliable operation of the equipment, the functional faults and difficulties ought to be eliminated, or at least brought down to a manageable level set subjectively by man;

g. The fault diagnosis and fault elimination processes involve a direct ·man-machine interaction;

Most mechanical, electro-mechanical, electro-chemical and chemo-mechanical systems are MPE. This contrasts with typical electronic systems, the states of which are binary.

1.2.2. Knowledge Base Description

(a) The RELSHELL knowledge base is structured as a fuzzy relation R from a set of influences I, acting on the equipment E studied, into a set of symptoms S of degradation of E. The sets S and I are represented by their respective frames F_0 (s_p) and F_0 (i_q), described in (34). The fuzzy membership value r (s,i) of R expresses the strength of the relationship holding between the influence F_0 (i_q) and the symptom F_0 (s_p) it generates. The r (s,i) is obtained by a mapping G from a lexicographic lattice, i.e. an ordered set of linguistic hedges, into the interval [0, 1]. Examples of commonly used linguistic hedges are "very strong", "weak", "non-existent", etc. Obviously, 1 and 0 represent very special values of the linguistic hedges, viz. an absolute necessity of occurrence of a phenomenon, resp. an absolute impossibility of its occurrence. In (34) it was stipulated that the mapping G should be a monomorphism.

b. Facts, situations or phenomena belonging to S are represented by one-dimensional zero-order hierarchical frames. The frames are formatted as follows: F_0 (s_p) = [Name of the p-th symptom; Symptom descriptor; Crisp threshold value activating the descriptor], where F_0 (s_p) is the zero-order frame representing symptom s_p. For example, $F_0(HT)$ = [High temperature; Temperature level; >70°C] indicates that there is a symptom called "high temperature" which is observed as soon as the value of its descriptor called "temperature level", exceeds 70°C. It means that below 70°C this particular frame will not be active in the knowledge base. The zero-order one-dimensional frames for I are formulated analogously.

c. The inference rules used in the RELSHELL are:

1. WHENEVER $(F_0 (i_q))$, THEN $(F_0 (s_p))$ becomes activated with a degree of possibility $r (s_p, i_q)$.

$$(1.2.2.1)$$

2. Q evokes $A(F)$

$$(1.2.2.2)$$

Rule (1.2.2.1) establishes the relation R. The condition WHENEVER (used instead of IF) expresses the time-dependence of $F_0 (i_q)$: the frame will become active as soon as the threshold value of its descriptor is exceeded. Rule (1.2.2.2) is used for handling fault diagnostic querries. The answer $A(F)$ to a query Q is a fuzzy quantity obtained from R by some fuzzy-logical or fuzzy-arithmetic operations.

d. Reasoning (or knowledge extraction) in the RELSHELL is done primarily by means of fuzzy-logical or fuzzy-arithmetic operations, e.g. by conditioning R by a fuzzy set having some useful properties required by the problem to be solved. An interaction operator, called iota, is used to condition R. In that manner, functions PRIOR(I) and LTS(I) were developed in (34).

1.3. RELSHELL - Extensions

The reasoning power of the RELSHELL can be enhanced by a more thorough deployment of the frames. The need for such an enhancement does not follow from the theory of ES - actually it would not be detectable by the theoretical AI-techniques available. Rather, it is dictated by the exigencies of the engineering situations which the RELSHELL has been designed to handle.

One such situation can be exemplified as follows. A macro-molecular component of a MPE is exposed to a variety of adverse influences. Let those influences, forming set I:{temperature, load, UV-light, chemicals}, give rise to some symptoms of deterioration of E, for instance S:{break down, opacity}. The elements of I and S can be represented by their respective one-dimensional, zero-order frames, with a specific $r(s_p, i_q)$ assigned to each WHENEVER-THEN rule. The situation will change substantially if more than one influence act on the component simultaneously. Thus, the subsets of I, e.g.{temperature, load}, {temperature, UV-light}, {UV-light, load, temperature}, etc, will each be characterized by different $r(s_p, i_q)$.

Under these conditions, the inference rule sub 1.2.2.1 can no longer hold. It has to be re-formulated as follows:

WHENEVER $\{F_{0,1}(i_1)$, and $F_{0,1}(i_2)$, and $\ldots$, $F_{0,1}(i_q)\}$, THEN $F_{0,1}(s_p)$ with strength $r(s_p; i_1, i_2, \ldots, i_q)$

$$(1.3.1)$$

Expression (1.3.1) represents a logical AND of various influences. It can be conveniently reduced to a single representative influence by defining higher-order frames:

DEFINITION 01: Hierarchical n-th order, k-dimensional frame
An n-th order, k-dimensional hierarchical frame $F_{n,k}(p)$ is a data
structure for representation of facts and/or phenomena (called par-
ameters p) formatted as follows:
  A. Zero-order k-dimensional frame
      $F_{0,k}(p)$=[Name of parameter p; Parameter descriptors DR_j; Crisp
              threshold values activating the descriptors]
  B. n-th order, k-dimensional frame
      $F_{n,k}(p)$=[$F_{n-1,k}(p)$; Fuzzifiers of p; Fuzzy values pertinent to
              each fuzzifier]

NOTE: Definition 01 is a generalization of the zero-order, one-dimensional
hierarchical frame, $F_{0,1}(p)$ (34). The dimensionality k of $F_{0,k}(p)$ is equal
to the cardinality of the sub-set of its descriptors. Consequently, a
multi-dimensional frame is active if, and only if, all of its descriptors
have been activated. In the continuation of this paper, only zero-order,
k-dimensional frames will be considered.

The reasoning rule (1.3.1) can now be written:
WHENEVER $F_{0,k}(i_q)$, THEN $F_{0,1}(s_p)$, with strength $r(s_p, i_q)$.

$$(1.3.2)$$

1.3.1. Example 1 - The composition of frames.   Let:

$F_{0,1}(i_1)$=[Component ageing; Time; XXX days]
$F_{0,3}(i_2)$=[Component deterioration; Time; Z days
                                UV-light; AA lux at λ_n,
                                Stress; KK N/mm2]
$F_{0,2}(i_3)$=[Component deterioration; Time; YY days,
                                UV-light; AA lux at λ_n]
$F_{0,1}(s_1)$=[Component failure; Component ageing; XXX days]
$F_{0,1}(s_2)$=[Component opacity; Age-dependent decrease in translu-
          cence; <BB lux at λ_n]

As an example, the following rules can be postulated, using (1.3.2):

$F_{0,1}(i_1)$ R $F_{0,1}(s_1)$=0.3
$F_{0,3}(i_2)$ R $F_{0,1}(s_2)$=0.5
$F_{0,2}(i_3)$ R $F_{0,1}(s_2)$=0.9
$F_{0,1}(i_1)$ R $F_{0,1}(s_2)$=0.2

The first rule can be verbalized as follows. Whenever the component ages
over a period of XXX days, then the possibility that the component will fail
in XXX days because of ageing is equal to 0.3.   Here, one-dimensional
frames are used.

    Rule two uses a three-dimensional frame of influences.   It reads:
Whenever UV-light of AA-lux at wavelength λ_n, and stress of KK N/mm² act
simultaneously on the component over a period of Z days, then the possibility is 0,5 that, because of ageing, the component opacity will reach a
degree when it lets through less than BB lux of light at λ_n .   The meaning
of the remaining rules can be interpreted in a similar manner.

    Example 1.3.1. shows that an enormous quantity of information can be
packed into a multi-dimensional frame, which results in increased flexibility of R. That is an advantage. The disadvantage is that the multi-
dimensional frames augment relation R considerably. The number of frames
increases because one and the same influence i_q can appear in several frames
of various dimensionalities. However, definition 01 implies that each frame

is active only if all of its descriptors have been activated. Therefore, the computational effort can be reduced by letting R disintegrate into modules consisting of the active frames only. This idea can be approached theoretically by studying the descriptors of a multidimensional frame.

Let $F_{0,k}(p)=[PP; DR; TV]$, where PP, DR and TV denote, respectively, the parameters of the data represented in the frame, the descriptors of the parameters, and the threshold values of the descriptors. The set of descriptors, DR, obviously has the following properties:

> The elements of DR, $\{d_j\}$, occur as:
> - singletons $\{d_i\}$,
> - subsets F_i of DR,
> - the whole set DR, whereby $\qquad\qquad\qquad\qquad$ (1.3.3)
> $\quad F_i \neq \emptyset$
> $\cup F_i = DR$.

Is there a suitable mathematical structure for the treatment of the multidimensional frames having the properties of the descriptors just outlined? There is; it is the matroid. (43)

DEFINITION 02: Matroid
A matroid on DR is called the pair $M=(DR, F)$, where $DR=\{d_1, d_2, \ldots, d_k\}$ is a finite set. F is a family of subsets of DR. The following conditions hold for a matroid:
1. $\{d_i\} \in F \qquad (i=1, 2, \ldots, k)$
2. $F \in F$, $F' \neq \emptyset$, $F' \subset F \Rightarrow F' \subset F$,
3. For each $S \subset DR$, the members that are maximal in S have the same cardinality.

The matroid was originally introduced to study axiomatically the properties of linear independence. The subsets $F' \subset F$ are the independent sets of M, whereas the subsets not belonging to F are called the dependent sets. In studying multidimensional frames, it is necessary to delimit the meaning of the matroid terminology within the framework of the RELSHELL.

PROPOSITION 03:

A zero-order multidimensional frame can be represented by a matroid. PROOF: A. For a multi-dimensional frame to be representable by a matroid, a unique correspondence between the elements of the matroid, and those of the descriptors of the frame, must be established. This can be done in the following manner. Let two sets, E1 and E2, have the same cardinality k. It is required that a one-to-one correspondence hold between the elements of E1 and E2. The correspondence is stipulated to be a bijective function β which has the following properties:

(a) it is surjective, i.e. each element $y \in E2$ is the image of at least one element $x \in E1$;
(b) it is injective, i.e. each element $y \in E2$ is the image of either one or no element $x \in E2$;
(c) being a function, the cardinality of β, $|\beta|=1$;
(d) for each bijection β from E1 to E2, there exists an inverse bijection β^{-1} from E2 to E1.
Thus, for $|E1|=|E2|$, the one-to-one correspondence is established uniquely.

B. Let E1 be subdivided into n subsets of cardinalities n_i, $i=1, 2, \ldots, k$. The pseudo-partition P, formed on E1, has the following properties:

(d) $\emptyset \in P$
(e) $\cup P = E1$
(f) $X, Y \in P \Rightarrow X \cap Y \in P$.
By virtue of A, a corresponding pseudo-partition is generated on E2.

C. Equate E1 with the matroid, and E2 with DR, the set of descriptors occurring in R. Thereby the vertices of the matroid become identical to the
elements of DR, and the (independent) subsets of the matroid form the frames
of R. The proof is thus completed.

1.3.2. Example 2

This example will illustrate the implications of definition 02 and
proposition 03. Let DR=$\{d_i\}$=$\{p, q, r, s, t\}$, and F=$\{\{p\}, \{q\}, \{r\}, \{s\}, \{t\},$
$\{p,q\}, \{p,r\}, \{p,s\}, \{p,t\}, \{q,r\}, \{q,s\}, \{q,t\}, \{p,q,r\}, \{p,q,t\}\}$. Is this a
matroid? To answer the question, the set DR must be analyzed in light of
definition 02.

Condition 1 of definition 02 is obviously fulfilled: all singletons
belong to the matroid.

Condition 2 stipulates that each subset of a subset of DR must belong
to F. To verify this requirement, commence with the subsets of cardinality
equal to 3. The subset$\{p,q,r\}$ contains the three singletons $\{p\}$, $\{q\}$, $\{r\}$
, and the subsets $\{p, q\}$, $\{p, r\}$ and $\{q, r\}$. These six subsets do belong
to F. A similar procedure applied to $\{p, q, s\}$ and $\{p, q, t\}$ also yields
positive results, which satisfies the condition.

Condition 3 is the most important one for deciding whether or not the
couple (DR, F) is a matroid. It all depends on the choice of S. To begin
with, choose S=DR. Then the maximal subsets $\{p,q,r\}$, $\{p,q,s\}$ and $p,q,t\}$ do
have the same cardinality equal to 3. Next, take S=$\{p,q,r,s\}$. The maximal
subsets of S belonging to F are $\{p,q,r\}$ and $\{p,q,s\}$; the condition 3 is thus
satisfied. Proceeding in the same manner, S can be taken to be
$\{p,q,r,t\}$, $\{p,r,s,t\}$, $\{p,r,s,t\}$, $\{q,r,s,t\}$, etc. Each of these choices satisfies condition 3, and (DR, F) is thus proved to be a matroid.

To show a counter-example, delete $\{p,q,r\}$ from F. Then, for
S=$\{p,q,r,s\}$, the subsets $\{p,q,s\}$ and $\{p,r\}$ are maximal, but they do not have
the same cardinality. Therefore, condition 3 is not satisfied and the family
of the subsets considered is not a matroid.

Further properties of the matroid, of interest in this paper, can now
be expounded.

DEFINITION 04: Rank of matroid.
The rank $r(S)$ of matroid M is defined to be $r(S) = \text{MAX}_{F \in F} |F_i \cap S|$

THEOREM 05

Let M be a matroid of rank $r(S)$. Then: $r(S) = \text{MAX } |S \cap F_i| \leq \text{MAX } |F_i|$.

PROOF: Let $|DR|$=n, and $|F_i|$ = j, j $\leq$ n. Two events are possible:
(a) S is identical to DR. Then, $F_i \subset S$. From definition 04 it then follows
that $r(S) = \text{MAX } |F_i|$;
(b) $S \subset$ DR. Then, $r(S) \leq |F_i|$.
Example 2 demonstrates some disadvantages involved in using the concept of
matroid, originating in the discrepancies between the postulates of the
theory, and reality: (1) The existence of a matroid on a set depends

largely on the choice of S and F.   (2) From the proof of proposition 03 it follows that the sets of the matroid (subsets of E1) correspond to subsets of descriptors, DR.   In reality some of these subsets (frames) perhaps do not exist, i.e. their relation $r(s,i)$ to the set of symptoms is empty, which increases the volume of R without doing any good.

To turn these disadvantages into advantages in a practical application, it would be good to get rid of the superfluous frames.   This can be accomplished by (a) substituting the whole set DR for S and (b) selecting each F_i so as to be identical to a distinct multi-dimensional frame $F_{0,k}(p)$. This approach leads to a mathematically simpler form of matroid, called the hypergraph (44). Before showing its practical applications, the basic properties of the hypergraph will be reviewed.

DEFINITION 06: Hypergraph.

If $M=(DR, F)$ is a matroid with rank $r(S)$, then the maximal independent sets of M form a uniform hypergraph of rank $r(DR)$.

DEFINITION: 07: Hypergraph

Let $X=\{x_1, x_2, \ldots, x_n\}$ be a finite set, and let $E=(E_i \mid i \subset I)$ be a family of subsets of X.   Suppose:
1. $E_i \neq \emptyset$
2. $\cup_i E_i = X$.
Then the couple $H=(X, E)$ is called a hypergraph consisting of vertices $x_i \in X$, and edges E_i.

DEFINITION 08: Simple hypergraph.

If the edges E_i of H are all distinct, then H is said to be a simple hypergraph.

DEFINITION 09: Rank of hypergraph.

Let $S \neq \emptyset$, and $S \subset X$.   Then $r(S)=MAX_i |S \cap E_i|$ is called the rank of H.

DEFINITION 10: k-section of hypergraph.

The k-section $H(k)$ of H is a simple hypergraph of rank $r_k(S)$, where $r_k(S)=\min |k, r(S)|$.

DEFINITION 11: Adjacent vertices and edges.

Two vertices of H are said to be adjacent if there is an edge E_i that contains both of these vertices.
Two edges of H are said to be adjacent if their intersection is not empty.

DEFINITION 12: Incidence matrix.

The incidence matrix of a hypergraph $H=(X, E)$ is a matrix $[a_{ij}]$ with m rows representing the edges of H, and n columns representing the vertices of H, such that

$$a_{ij} = \begin{cases} 1 & \text{if } x_j \in E_i \\ 0 & \text{if } x_j \bar{\in} E_i \end{cases}$$

DEFINITION 13: Dual of a hypergraph.

A hypergraph $H'=(E; X_1, X_2, \ldots, X_n)$ whose vertices are points e_1, e_2, e_3, $\ldots$, e_m (that represent E_1, E_2, $\ldots$, E_m) and whose edges are sets X_1, X_2, $\ldots$, X_n (that represent x_1, x_2, $\ldots$, x_n), where, for all j, $X_i=\{e_i | i \leq m, x_j \in E_i\}$, is called the dual of hypergraph H. From definitions 12 and 13 it follows that the incidence matrix of H' is the transpose of the incidence matrix of H.

DEFINITION 14: Clique of rank h.

Let $H=(X, E)$ be a hypergraph of rank r, and let $h \leq r$. A set $A \subset X$ is defined to be a clique of rank h if either $|A| < h$, or $|A| \geq h$ and each subset of A with cardinality h is contained in at least one edge of H.

DEFINITION 15: Representative graph of H.

Let $H=(X; E_1, E_2, \ldots, E_n)$ be a hypergraph with n edges. The representative graph of H is defined to be a simple graph G of order n whose vertices x_1, x_2, $\ldots$, x_n, respectively, represent the edges E_1, E_2, $\ldots$, E_n of H. Vertices x_i and x_j are joined by an edge if, and only if $E_i \cap E_j \neq \emptyset$.

DEFINITION 16: Power of E_i.

Let E_i be an edge of H. Then $P(E_i)$, the power of E_i, is equal to $|\cup_j (E_i \cap E_j)|$ for all i,j=1, 2, $\ldots$, n, such that $i \neq j$. MAX $P(E_i)$ identifies the most powerful edge of H.

DEFINITION 17: Solitary vertex and isolated vertex

Let x_v belong to E_v, such that $|E_v| > 1$. Then x_v is called a solitary vertex if, and only if, $x_v \bar{\in} E_v \cap E_j$ for all positive integers j. If $|E_v|=1$, and $E_v \cap E_j=\emptyset$ for all positive integers j, then x_v is called the isolated vertex.

DEFINITION 18: Isolated edge.

An edge E_i, such that $E_i \cap E_j=\emptyset$ for all j=1, $\ldots$, n is called the isolated edge.

It is not difficult to apply these graph-theoretic notions to the knowledge base studied. The vertices of the hypergraph correspond to the elements of DR. Its edges, some subsets of DR, correspond to the active (in some instant) multi-dimensional frames of R, except in the case of an isolated vertex which represents a one-dimensional frame. The frames are activated by the threshold values of their descriptors. To ascertain which frames are active, given that k descriptors of DR are active, use theorem 19.

THEOREM 19:

Let there be k active descriptors of frames in a knowledge base R, represented by H. Let k be the k-section of H. Then, a frame $F_{0,k}(p)$ of R, corresponding to E_i of H, is active if and only if $k \geq r(S)$.

PROOF: Suppose the stipulation is not true, i.e. $k < r(S)$. From theorem 05 and definition 09 it follows that $r(S)$ cannot be larger than $|E_i|$, provided E_i is presumed active and S=DR. By definition 10, $r_k(S)$ cannot be larger than $r(S)$; it can be smaller. However, $r_k(S) < r(S)$ implies that $k < r(S)$. In that case, $k<|E_i|$ and E_i cannot be active, because some of its descriptors remain inactive. This contradicts the stipulations of the theorem.

In an automatically monitored equipment, the k could be critical levels of signals from sensors, which activate some subsets of DR with cardinality $\leq k$, and thus all the frames containing those subsets of DR. The module(s) consisting of the active frames is visualised by the representative graph of H. Unless redundancies are present, the hypergraph is always simple.

2. SAMPLE CALCULATION

A specific feature of knowledge-based fault diagnosis is the fact that usually the number of querries the user addresses to the knowledge base is limited, but the answers to those querries must be relevant, repeatable, precise and compatible with experimental results. In a fuzzy knowledge base of the RELSHELL type, such answers result from fuzzy-logical and fuzzy-mathematical operations. In general, the type and number of the operations performed depend on the specific requirements of a given problem. The sample calculation which follows illustrates, in a step-by-step manner, the use of the theoretic concepts developed in this paper, as well as in (34). The data used have been taken from a real project: fault analysis of a high-speed electro-mechanical device. The original knowledge base contains almost 200 rules, but only a few of them are shown in Table 1 for demonstration of the computational procedures.

1. Start by examining and listing the influences acting on the equipment, as well as the symptoms of deterioration of the equipment. These lists can be compiled using a variety of sources, such as engineering intuition and experience, results of engineering bench tests, manufacturing shop-floor experience, feed-back from the users and the customer service, engineering specifications and drawings, or specialized and trade literature (45-46).

Suppose the influences and the symptoms to be used in this sample calculation, along with their descriptors and threshold values formatted into frames, are:

FRAMES - I:

$F_{0,1}$ (i_1) = [Dimension P; Thickness; <4mm]
$F_{0,1}$ (i_2) = [Dimension W; Δ-length; >0.1mm]
$F_{0,1}$ (i_3) = [Clearance between elements Q; Distance variation;
                    <0.05 or >0.1mm]
$F_{0,1}$ (i_4) = [Cam wear; Erosion depth; >2mm]
$F_{0,3}$ (i_5) = [Vibrations; Cam wear/erosion; >2mm;
                    Finish roughness; >0.1mm
                    Gear run-out; >0.5mm]
$F_{0,5}$ (i_6) = [Assembly accuracy; Cam erosion depth; >2mm;
                    Plate length var.; >0.5mm
                    Spacer thickness var.; >0.2mm
                    Elements clearance var.; <0.5 mm, or>0.1mm]
$F_{0,4}$ (i_7) = [Lubrication failure; Defficient oil flow; <1g/minute;
                    Defficient pad bonding; Loose;
                    Wick contact broken; Loose;
                    Elevated temperature; >75°C]

FRAMES - S:
$F_{0,1}$ (s_1) = [Non-uniform movement; Speed variation; >0.1m/s

Table 1

			SYMPTOMS				
			s_1	s_2	s_3	s_4	s_5
			DS	DA	TQ	ET	CW
I N F L U E N C E S	i_1	DP	0.6		0.7	0.5	
	i_2	DW	0.4		0.7	0.7	0.8
	i_3	CQ	0.7		0.7	0.8	0.7
	i_4	CW	0.7	0.2	0.7	0.2	0.5
	i_5	CW FR GR	0.7		0.5	0.7	0.4
	i_6	CW PF PL ST CQ	0.9	0.5	0.9	0.9	0.8
	i_7	OF PB WC ET	0.9		0.9	1.0	0.7

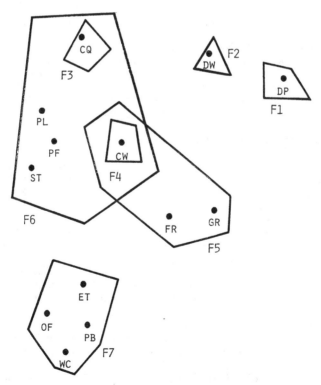

Fig. 1 Hypergraph representation of the knowledge base used in sample calculation.

$F_{0,1}$ (s_2) = [Debris; Accumulation; >1mg/8h;]
$F_{0,1}$ (s_3) = [Excessive torque; Torque; >3 Nm;]
$F_{0,1}$ (s_4) = [Elevated temperature; Temperature; >75°C]
$F_{0,1}$ (s_5) = [Cam wear; Erosion depth; >2mm]

2. Next, establish the fuzzy relation holding between the elements of the sets I and S. It was shown in (34) that this relation can be found as a mapping from a set of linguistic hedges into [0, 1]. Moreover, that mapping has to be represented by a monomorphism. These requirements are easy to meet. Start by listing the linguistic hedges, i.e. some adjectives qualifying the strength of the relationship existing between each pair of influence and symptom frames. The qualifying adjectives can be words such as a "weak", "non-existent", "strong", highlighted adverbs like "very", "medium", "almost", etc. For instance, in table 1, the relationship between $F_{0,5}$ (i_6) and $F_{0,1}$(s_3) is "very strong". Suppose the scale of linguistic hedges stretches from "non-existent" at the low end to "total" at the high end, with "medium strong" in the middle. Finer nuances (e.g. "very weak", "very strong", "more than medium") occupy the intermediate positions in the lattice.

Now, establish the numerical scale. This is easy, because the scale is the interval [0, 1], into which the linguistic hedges can be mapped. To be a monomorphism, the mapping must preserve the order (hierarchy) of the linguistic hedges, i.e. a higher-level hedge must be linked with a higher numerical value on the 0-to-1 scale. Therefore, assign the strength 0 to the hedge "non-existent", the strength 1 to the hedge "total", and the strength 0.5 to "medium". The intermediate hedges correspond to the remaining numerical values. The fuzzy knowledge base R thus obtained is shown in table 1.

3. Finally, use the knowledge base to extract fault diagnosis knowledge from R. As the RELSHELL is based on fuzzy arithmetics, the responses to the querries posed will be calculated and presented primarily in numerical form. A verbal commentary to the results is a trivial matter of the user interface, and will not be handled here. Although an unlimited number of mathematical structures for fuzzy reasoning can be generated, usually merely a small number of them are interesting. The most relevant querries can be subdivided into the following categories:

A. Querries related to zero order frames

A.a. Given an influence frame $F_{0,k}$ (i_q), which symptoms $F_{0,1}$ (s_p) can be anticipated? The answer can be found by applying inference rule (1.2.2.1), and enumerate the symptom frames it generates. Example: $F_{0,5}$ (i_6) generates: {s_1, s_2, s_3, s_4, s_5}.
A.b. Given an influence frame $F_{0,k}$ (i_q), which is the most prominent symptom of equipment deterioration it brings about by the influence frame? To answer the query, find the first projection over R. For instance, for $F_{0,1}$ (i_1), it is $F_{0,1}$ (s_3)=0.7.
A.c. Given some influence frame $F_{0,k}$ (i_q), list the symptoms whose importance exceeds some threshold level α. That is, seek α≤r (s_p, i_q). Example: let α=0.5. Then, for $F_{0,1}$ (i_2), the symptoms exceeding the level α= 0.5 are {s_3, s_4, s_5}.
A.d. Given set I, let i_k, i_l be two of its elements. Which of them is more important? To answer the query, use the relation developed in (34) which stipulates that i_k is more important than i_l if $\Sigma'F_{0,k}(i_k) > \Sigma'F_{0,k}(i_l)$, where $\Sigma'F_{0,k}(i_k) = \Sigma_{p=1}^{p} r(s_p, i_q)/|\Gamma_R|$. $\Sigma'F_{0,k}$ (i_k) is the cardinality of i_k as defined in Kaufmann, (36), divided by the cardinality of its support. This approach makes it possible to distinguish between sets having the same

cardinality, but very different fuzzy membership values of their elements. Compare, for instance, fuzzy sets {0, 0, 0, 0.8, 0} and {0.1, 0.2, 0.2, 0.3}. The former set has only one element belonging to the support, but that element has a large fuzzy value. The latter set consists of a large number of elements of small values. Therefore, the former set can have a much greater impact on the degradation of the equipment. Example: for $F_{0,1}(i_4)$, $\Sigma'\ F_{0,1}(i_4)=(0.7+0.2+0.7+0.2+0.5)/5=0.46$, whereas for $F_{0,3}$ (i_5), Σ' $F_{0,3}(i_5)=(0.7+0+0.5+0.7+0.4)/4 = 0.575$. Therefore, $F_{0,3}$ (i_5) is slightly more important than $F_{0,1}(i_4)$.

B. Querries related to higher order frames

In (34) two functions depending on higher order frames were shown. These were obtained by fuzzifying a zero order frame.

B.a. The function called $PRIOR(I|s_p)$ ranks the influences (first order frames), bringing about symptoms s_p in the order of their magnitude. The higher order of the frame results from conditioning R by I, where I is a fuzzy set expressing the ease with which an influence can be controlled by man. An influence which is more difficult to control will have a low value of "ease", and v.v. The function PRIOR (34) is formulated as follows: $PRIOR(I)=\downarrow(R \iota I)$, where ι was shown to materialize as an element-by-element exponentiation of R by I over all s_p belonging to R . For a given symptom frame, PRIOR can be re-written as:
$PRIOR(I|s_p)=\downarrow(F_{0,k} (s_p) \iota F_{0,1} (i'))$.
For example, suppose I ={0.7, 0.7, 0.8, 0.2, 0.3, 0.5, 0.1}. Then, taking $F_{0,1} (s_1)$, $PRIOR(I'|s_1)=\downarrow(0.6exp0.7, 0.4exp0.7, 0.7exp0.8, 0.7exp0.2, 0.7exp0.3, 0.9exp0.5, 0.9exp0.1)=\downarrow(0.69, 0.53, 0.75, 0.93, 0.89, 0.94, 0.98)$ $=>\{i_7, i_6, i_4, i_5, i_3, i_1, i_2\}$.

B.b. Another function, called Long-Term-Stability (LTS), expresses the invariance of the equipment over a period of time. (LTS(I) is defined in (34) to be: $LTS(I)=\downarrow(PRIOR(I)\iota|I')$, where I' expresses the experimenter's subjective confidence that the state of the equipment will not change in a given time period. Again, a large value of I' indicates that the frame to which it is assigned contributes little to the variability of the state of the equipment during the period given. The structure of LTS is similar to that of PRIOR. However, the LTS yields results which express the more unfavorable combination of the ease of handling of the influences, and the invariability of the state of the equipment in time. Because of the properties of ι and the structure of the two functions, the order in which I and I' are applied is irrelevant.

As an example, suppose the LTS of frame $F_{0,1} (s_1)$ is sought, given that I'={0.5, 0.5, 0.3, 0.5, 0.7, 0.8, 0.4}. Then, $LTS(I'|s_1)=\downarrow(PRIOR(i_q)|(s_1 \iota I'$)$)=\downarrow\{0.83, 0.72, 0.91, 0.96, 0.92, 0.95, 0.99\}$, i.e.$\{i_7, i_4, i_6, i_5, i_3, i_1, i_2\}$. This is the ranking of the influences by the rate of their contribution to the variability of s_1 within a given period. The most interesting influence seems to be i_7: it is most difficult to handle, and it is the greatest contributor to the variability s_1.

C. Querries related to the transpose of R

Essentially the same operations can be performed as shown sub A and B above. This time, the symptom is given, and influences are sought.

D. Simulation of equipment improvements

The database is querried about changes induced by varying the fuzzy value of the relationship between the influence and symptom frames. The variations of r (s_p, i_q) follow from some engineering modifications of the equipment studied. The goal is to determine whether a modification is worthwhile, considering the costs, effort, and engineering or social problems involved.

E. Modularization of the influence frames

Modularization of the knowledge base, in the context of this paper, means quasi-partitioning of the set of influence frames into subset of frames that are, under given operational conditions, active, resp. inactive. In a small knowledge base, like the one shown in table 1, modularization lacks any practical importance. However, in large knowledge bases, modularization helps in at least three ways. First, it provides information on which frames are active under given physical conditions. Thereby it also provides information on the severity of the influences acting on the equipment. For instance, the descriptor called "cam wear" appears in influences i_4, i_5 and i_6. However, its effect on the state of the equipment is different each time. While in i_4 it is the sole influence, it makes its frame active as soon as the erosion depth exceeds the value specified. In i_5 5 and i_6 "cam wear" is a descriptor in 3-, resp. 5-dimensional frames, each having a very different r (s_p, i_q).

Secondly, modularization helps to identify the most critical influences. Displaying the k-dimensional frames in the form of a hypergraph makes it possible to find the most powerful edge of H, thus making out the most critical configuration of influences. For instance, in this sample calculation, $F_{0,5}$ (i_6) is the most powerful edge, with $P(F_{0,5}(i_6))=4$. Vertices PF (plate flatness) and LV (plate length variation) of $F_{0,5}$ (i_6) are solitary vertices, i.e. as descriptors they will never affect another frame. Vertex DL (length variation) of $F_{0,1}$ (i_2) is an isolated vertex, i.e. it can never be affected by another frame. Likewise, $F_{0,4}$ (i_7) is an isolated edge and therefore it constitutes an independent module of R. Figure 1 shows the hypergraph representing the database used for this sample calculation.

Thirdly, modularization minimizes the risk of obtaining erroneous results following from frames update if the elimination of the old frames is omitted. Then, two situations are possible. Either the new frame specifies an r (s_p, i_q) which is numerically close to the frame it has replaced, or the two values of r (s_p, i_q) are very different. Because of the ranking operator ↓ deployed in PRIOR and LTS, in the former case the old and the new frames will lie close to each other, without introducing any substantial error into the conclusions drawn. The latter situation will move the more severe influences to the front, thus highlighting the more severe condition.

REFERENCES

1. A. A. Maslow, *Motivation and Personality*, Harper & Brothers, New York, 1954.

2. F. Herzberg, One more time: How do you motivate employees?, *Harvard Business Review*, Jan-Feb. (1968).

3. D. MacGregor, *The Human Side of Enterprise*, McGraw-Hill, New York, (1960).

4. W. W. Danial, What interests a worker, *New Society* (March, 1972).

5. A. Kelly, Topics in Terotechnology: An Instructional Series, 10, Motivation of the Maintenance Tradeforce, *Maint. Mgmt. Intl.* 4(2) 71-81, (1984).

6. K. D. Duncan and M.J. Gray, An Evaluation of a fault-finding training course for refinery process operators, *J. Occup. Psychology, 48,* 199-218 (1975).

7. W.B. Rouse, R.M. Hunt, in *Advances in Man-Machine Systems Research* (W. B. Rouse, ed.), Vol. 1, JAI Press, Greenwich, CT, (1983).

8. L. F. Pau, *Diagnostic des pannes: approche par la reconnaissance des formes,* Cepadues Editions, Toulouse (1975).

9. I. Gazdik, Stimulation Enhances Inventiveness on the Shop-Floor, *Technovation 4,* 131-141 (1986).

10. A. Barr and E. A. Feigenbaum (eds), *The Handbook of Artificial Intelligence,* Vol. 1 and 2, Kaufmann, Los Altos (1982).

11. P. R. Cohen and E. A. Feigenbaum (eds), *The Handbook of Artificial Intelligence,* Vol. 3, Kaufmann, Los Altos (1984).

12. B. G. Buchanan and R. O. Duda, Principles of Rule-Based Expert Systems, Fairchild Technical Report No 626 (1982).

13. B. Chandrasekaran, Generic tasks in knowledge-based reasoning: High-level building blocks for expert system design, *IEEE Expert, 1(3),* 23-30 (1986).

14. R. Davis and D. Lenat, *Knowledge-Based Systems in Artificial Intelligence,* McGraw-Hill Computer Science Series, New York (1982).

15. H. Falk, Design Tools Combine Expert and Algorithmic Software. *Comp. Des.,* (Sept. 15, 1986).

16. R. Fikes and T. Kehler, The role of frame-based representation in reasoning, *Comm. ACM, 28(9),* 921-932 (1982).

17. B. R. Gaines and M. L. G. Shaw, Induction of Inference rules for expert systems, *FSS 18(3),*315-328 (1986).

18. D. Gregory, Delimiting Expert Systems, *IEEE Trans. SMC 16(6),* 834-843 (1986).

19. P. Haley and C. Williams, Expert system development requires knowledge engineering, *Comp. Des., 25(4),* 83-88 (1986).

20. F. Hayes-Roth, The knowledge-based expert system: a tutorial, *IEEE Computer,* (September 1984), p.11-28.

21. F. Hayes-Roth, Rule-based systems, *Comm. ACM 28(9),* 921-923 (1985).

22. F. Hayes-Roth, The machine as a partner of the new professional, *IEEE Spectrum 21(6),* 28-31 (1984).

23. T. Jupille, Expert systems are new textbooks, *Res. and Dev. 28(12),* 52-55 (1986).

24. K. L. McGraw, Guidelines for producing documentation for expert systems, *IEEE Trans. Prof. Com., PC-29(4),* 42-47 (1986).

25. C. V. Negoita, *Expert Systems and Fuzzy Systems*, The Benjamin/Cummings Publ. Corp., Menlo Park (1985).

26. K. Niwa, A knowledge-based human computer cooperative system for ill-structured management domains, *IEEE Trans. SMC-16(3)*, 335-342 (1986).

27. P. Politakis, *Empirical Analysis for Expert Systems*, Kaufmann, Los Altos (1986).

28. B. G. Silverman and A. G. Tsolaki, Expert fault-diagnosis under human-reporting bias, *IEEE Trans. Rel. R-34(4)*, 366-373 (1985).

29. W. E. Snydam Jr., AI becomes the soul of the new machines, *Comp. Des.* 25(4), 55-70 (1986).

30. M. Stefik, J. Aikins, R. Balzer, J. Benoit, L.Birnbaum, F. Hayes-Roth and E. Sacerdoti, The Organization of Expert Systems: A Prescriptive Tutorial, XEROX Palo Alto Research Centers, VLSI-82-1 (January, 1982), p. 65.

31. V. R. Waldron, Interviewing for knowledge, *IEEE Trans. Prof. Comm.* PC-29(2), 31-34 (1986).

32. P. Wallich, Software "doctor" prescribes remedies, *IEEE Spectrum 10,* 43-48 (1986).

33. P. E. Lehner, On the role of artificial intelligence in command and control, *IEEE Trans. SMC 16(6)*, 824-833 (1986).

34. I. Gazdik, in *Fault Diagnosis, Reliability and Related Knowledge-Based Approaches* (S. Tzafestas, M. Singh and G. Schmidt, eds.), Reidel, Dordrecht (1987).

35. A. Jones, A. Kaufmann and H-J. Zimmermann, *Fuzzy Sets Theory and Applications, NATO ASI-Series,* D. Reidel, Dordrecht (1986).

36. A. Kaufmann, *Introduction a la theorie des sous-ensembles flous,* Tomes I-IV, Masson, Paris (1975-1977).

37. W. Karwowski and A. Mital (eds.), *Applications of Fuzzy Set Theory in Human Factors,* Elsevier, Amsterdam (1986).

38. L. A. Zadeh, Fuzzy logic and approximate reasoning, *Synthese 30,* 407-428 (1975).

39. L. A. Zadeh, The concept of linguistic variable and its application to approximate reasoning, Part 1, *Inf. Sci. 8,* 199-249 (1975); Part 2, *Inf. Sci. 8,* 301-357 (1975); Part 3, *Inf. Sci. 9,* 43-80, (1976).

40. L. A. Zadeh, Fuzzy Logic and Approximate Reasoning, *Synthese 30,* 407-428 (1975).

41. L. A. Zadeh, in *Advances in Fuzzy Sets Theory and Applications* (M. Gupta, R. Ragade and R. Yager, (eds.), pp. 3-18, North Holland, Amsterdam (1979).

42. L. A. Zadeh, The role of fuzzy logic in the management of uncertainty in expert systems, *FSS 11,* 199-227 (1983).

43. C. P. Bruter, *Les Matroides - Nouvel Outil Mathematique,* Dunod, Paris (1970).

44. C. Berge, *Graphs and Hypergraphs,* North-Holland, Amsterdam (1979).

45. L. F. Pau, *Failure Diagnosis and Performance Monitoring,* Marcel Dekker, (1981).

46. H. P. Rossmanith (ed.), *Structural Failure, Product Liability and Technical Insurance,* North-Holland, Amsterdam (1984).

LIST OF SYMBOLS

$A, B, \ldots, Y, Z$	sets
$a, b, \ldots, y, z$	elements of sets
$A, B, \ldots, Y, Z$	fuzzy sets
$a, b, \ldots, y, z$	fuzzy membership functions
DR	data descriptors in frame F
E	equipment studied
$F_{n,k}(p)$	n-th order, k-dimensional hierarchical frame
G	mapping from set to set
H	hypergraph
I	set of influences acting on equipment E
I	set of positive integers
M	matroid
PP	data parameters in frame F
R	fuzzy relation from I to S
$r(s_p, i_q)$	fuzzy membership value of R at s_p, i_q
$r(S)$	rank of matroid M
S	set of symptoms of degradation of E
TV	threshold values of descriptors in frame F
α	level of a fuzzy set
ι	aggregation operator
Γ_R	support of R
λ	wavelength
$|\bullet|$	cardinality of $\bullet$
$\downarrow$	ranking operator

5

A SPECIFICATION METHOD FOR DIAGNOSTIC SYSTEMS

IN LARGE INDUSTRIAL ENVIRONMENTS

Jean Maroldt

CRIN: Centre Recherche en Informatique de Nancy
Vandoeuvre-lès-Nancy, France

1. INTRODUCTION

Fault diagnosis in large industrial environments allows to draw up conclusions about the behaviour of the plant. These conclusions point out whether there are one or several components which are out of order, broken or unsettled.

To set up such conclusions a diagnostic system uses a knowledge base in which concepts are mostly expressed in terms of inference rules, semantic nets, object oriented languages or a composition of these three formalisms[1,2]. At present time only a few tools help the system designer to create a prototype kernel and then to develop the operational system integrated on the industrial site with the same tools.

We first describe the specific characteristics of fault diagnosis in large industrial plants, then we analyze two categories of approaches used to design diagnostic systems. The development of such a system requires a global approach strongly linked to the maintenance staff experiences. The specification tools must be as general as possible in order to be suitable for different application types. These tools may be grouped in two levels:

— unambiguous concepts to collect the knowledge and express the different expertise levels

— a method and a set of programs which integrate and handle the successive knowledge layers.

Our objective is to propose a method and a set of formal tools satisfying such requirements of diagnostic system design.

2. SPECIFIC TROUBLESHOOTING FEATURES IN INDUSTRIAL ENVIRONMENTS

A number of features are characteristic of troubleshooting in large industrial plants. Among them, we can quote:

— The complexity and diversity of such installations with sophisticated equipment; their maintenance requires highly specialised experts from different areas.

- The need for different staff expertises (mechanics, automata, computer sciences, etc).

- A periodical updating of equipping traditional plants with new technologies. The integration of new equipment in older installations increases heterogeneity and complexity.

- A computerized process, a function of which is to supervise the equipment and to treat some defects. It delivers a lot of data and even this process may be itself faulty.

- A continuous running process, for technical or economical reasons it is often impossible to bring the production process to a complete halt and therefore the maintenance activities are conducted on equipment which is kept running or brought to a slow running pace.

2.1. The troubleshooting problem

The heart of the problem is that no necessary obvious link exists between the delivered message and the component's failure causing it. Moreover, the messages are elaborated at a high hierarchical level and result from the automata fault diagnosis.

Not all the devices are controlled; in particular transmission components such as relays, commutators or connections can be faulty. A first task is to verify the correct behaviour of the computer system, so that the diagnostic system may rely on the delivered messages.

A defect may cause other defects, so visible symptoms must be interpreted in terms of original or subsequent observations. Another important features is that a defect has to be interpreted in a given context. An alarm may be of no significance in one case or may be the symptom of a component failure in another context.

2.2. The fault detection

The maintenance technician finds rapidly the defect when he has been given a good hint. For that he uses his own experience and intuition and rarely analyzes the component in detail to develop a deep reasoning. Fairly often he relies on analogies.

Usually he first verifies the good running of equipment that is simple to be diangosed and close to him, even if though the gathered information is not very instructive. Sometimes, he does not consider the fault context because he does not analyze information on a long period. Moreover, he is inclined to believe too quickly that he has found the defect cause and he tries to convience himself that the defection observed symptoms are those he would like to find. As a consequence, he gives up instead of reconsidering the whole context.

2.3. The Analyzed Plant Structure

To illustrate our diagnostic system design method, we present in this paper several faults diagnosed on a large continuous running plant: a blast furnace. We can not describe such a complex installation in a few lines, but we want to point out some important features. A blast furnace includes a lot of various components which are widely spread on the installation: some gas analyzers may be several hundred meters away from the sampling place; other components are on very high and dangerous access (CO has, high temperature) and so on. The plant is a continuously running process with permanent coke and ore loading. It never comes to a halt but it may be brought to a slow running pace.

When designing a diagnostic system it is absolutely necessary to take into account the plant hierarchy[3]. The functional components of a blast furnace are the following:
   a) the computer system: the running of the furnace is conducted by
      a central process-computer whose principal functions are:
         i) the computation of the furnace simulation mathematical
            model
        ii) the management of local efficiencies and global produc-
            tion
       iii) the identification and correction of some defects
   b) the furnace loading: coke and ore loading;
   c) the furnace gas analysis: furnace output gas analysis;
   d) instrumentation: sensors, analyzers, etc.;
   e) the furnace process: it is defined by all the chemical, thermal
      or mechanical processes involded in the cast iron production;
   f) the cowpers: thermal refulation of the furnace.

It should be noticed that a functional component is a technological entity not necessarly located at the same place. The conclusions of the mathematical model are symbolized by two variables:

W_e: It characterizes the blast furnace thermal balance; it defines the heat quantity necessary to produce cast iron and dross brought to the production of one ton cast iron (therms/cast iron produced); this thermal balance is essentially computed from the introduced ore and coke quantity and the producted cast iron tons; the ore, coke and cast iron quantities are determined by complex weighting instrumentations.

The measurements through the analyzers must be very accurate because 1% deviation of the sensors means 5-10% divergency on the cast iron production.

3. RELATED WORKS

In the design of diagnostic systems two quite different approaches usually occur:
   — the heuristic approach
   — the fundamental approach.

In the experimental approach of diagnostic reasoning, also called *diagnosis from experience*, heuristic information plays a dominant role. The diagnostic knowledge is expressed using heuristic rules and statistical intuitions given by domain experts. The plant structure is not, or only to a slight extent, taken into account.

Under the second approach, often referred to as *"deep"* (*fundamental*) *knowledge approach*, one begins with the description of the system (a physical device or a real world setting) together with the observation of its behaviour. The only available information is the description of the system, i.e., its design or its structure and the observation of its behaviour. In particular, no heuristic information about system failures is available.

3.1. The Heuristic Approach

A well known example of this approach is MYCIN[4]. Another more recent example is KADS (Knowledge Acquisition and Documentation Structuring)[5,6]. This method is strongly structured and defines four levels for the human expert knowledge:

— the domain level: concepts, relations, structures

— the inference level: metaclasses, inference rules

— the task level: goal tasks (depending on the application type)

— the strategy level: plans, meta-rules, repairs, impasses.

The concepts (e.g. symptom, fault index, etc.) used in KADS have no formal definition. Therefore a weakness in this approach is that no strong semantics are imposed. Moreover, KADS is a general knowledge acquisition tool. For example, it was used in the field of financial loans and seems to be of less practicallity for fault diagnosis than a method taking into account the industrial environment specific features[7].

3.2. The Fundamental Approach

This more universal approach is illustrated by two different attempts:

— qualitative physics

— first principles

3.2.1. Qualitative Physics

The qualitative physics objective is to understand the physicist's reasoning modes at a state where the numerical aspects of the problem are ignored. In many cases the discipline is understood through the knowledge of physics[10,11] (e.g. knowledge of the principles which govern an intergrated electronic circuit and application of these principles to the circuits to be diagnosed). In particular, qualitative physics allow finding through experiments the conceptual or production defects that frequently cause faults.

3.2.2. The First Principles

Essentially the first principles[12,13] require a formal system description. Then the theory leads to an algorithm for computing all diagnoses and various results concerning principles of measurement. The discrimination of competing diagnoses is derived from this computation. In this framework diagnoses are interpreted as conjuctions of certain faulty or abnormal components merged in normal ones.

If the system is defined in first order logic, then the diagnosis appeals to a satisfiability check for such arbitrary formulae. Since there is no decision procedure for determining the satisfiability of first order formulae, diagnosis cannot be computed in the most general case. Nevertheless there exist practical settings where consistency is decidable, hence diagnoses are computable (e.g. switching circuits in a binay full adder).

Qualitative physics and first principles are both interesting ways

to run a diagnostic process[4]. Unfortunately, it is impossible to give
now a complete formal description of a large industrial plant. For that
reason, we suggest an intermediate approach using heuristic knowledge[15]
for fault diagnosis on peripheral equipment (automata, instrumentation,
etc.) and a more fundamental knowledge to diagnose faults occuring on
functional components directly involved in the central production process.

4. EXPERT KNOWLEDGE ACQUISITION

For the knowledge base development the system designers usually analyze some faults and point out the related elementary symptoms. They have
then to deal with a *breadth problem* (the application field definition)
and a *depth problem* (elementary symptom definition). For instance they
have to decide whether the elementary symptoms is the message determined
by the central computer (e.g. sensors out of service) or is the verification with a voltmeter indicating that there is no more voltage. The system designer has to decide the level up to where he has to analyze the
symptom (for example, by automaton checking).

To avoid a dangerous mixture of different problems, we propose the
following steps[16]:

— define a fault list which the system has to diagnose (definition
  of the application field);

— specify the context in which fault symptoms are valid. Industrial
  equipment shows different running states according to the context
  in which it operates. For example, when it rains, H_2 rate increases abnormally in the furnace gas but there is no defect;

— build a list of all the necessary symptoms for the diagnosis of
  each fault index. A fault may induce several symptoms not included in the application field. These complementary symptoms
  must be examined without analyzing the faults outside the application field. Otherwise the application field risks to be enlarged
  to all possible faults of the plant;

— include for some symptoms their time evolution, i.e. the history
  of these symptoms;

— verify that the useful fault symptoms are correctly transmitted
  from the sensors until their display on screens or printers. This
  transmission is controlled and supervised by automata, and the
  behaviour of these devices must be included in the context of
  the analyzed faults.

Uncertainty factors are often used in these systems for two different
reasons: (a) the collected information about the plant behaviour is not
certain, (b) the execution of the diagnostic system has to be controlled.
In our method the message validation, the ability of knowledge structuring,
and the use of two levels in the design process avoid these uncertainty
factors, often difficult to evaluate.

Now, we shall define the knowledge base using two conceptual levels,
as the human experts do, an *object level* and a *control level*[17].

4.1. The Object Level

At this level the system designer introduces the heuristic knowledge
for diagnosing the environmental equipment fault indices. The goal is to
allow the maintenance staff to induce, from context and symptoms, a list
of fault indices. The symptoms involved are usually of lower level and

their significance is well established. In addition they are very often directly accessible in the central control room.

4.2. The Control Level

At he control level, the designer includes the strategies to determine the final solutions. These strategies are efficient only if they rely on a formal model which simulates the industrial plant process.

The hypotheses of the control level are fault index classes, output of the object level, or former conclusions of strategy reasoning. In our method, we define two kinds of classes. Some classes are specific to each functional component and defined at the object level. There is one general class, which hypotheses come from different components. This class allows the analysis of indices related to the whole plant.

To determine the list of effective faults, the control level:

— proposes the verification of the variables elaborated by the mathematical model,

— suggests long run, difficult or dangerous symptoms analyses,

— restricts the investigation field (e.g. convergence to one fault) or changes the investigation field when a context change occurs.

It is very important for the designer to propose verification tests on the plant for confirming or rejecting contexts or fault indices elaborated by the object level. When operations are suggested by the system, the ergonomic factor is important. Some verifications can be done in a short time (e.g., automaton verification) whereas others need long time to perform (e.g., verification of sub-parts far from the maintenance staff) or they are dangerous (instrumentation verification at the top of the blast furnace, where the gas mask is needed).

5. OBJECT AND CONCEPT DEFINITIONS

5.1. The Objects

In fault diagnosis some notions are commonly used, namely symptom, fault index, history, etc. However very often, the interpretation of these notions changes according to the experts. To formalize these notions and express their semantics, we use formal logic.

Example:

Consider a business bulding with n offices and a fire detector in each room. The detector indicates fire alarm if the temperature is higher than $70^{o}C$. In the control room each detector is connected to its own alarm signal. The corresponding expert rule is expressed by the following formula:

$$(\forall N:\text{fire\_alarm } (N):-(\exists T:\text{detector } (N,T), (T>70)))$$

In this example full first order logic is necessary to avoid the definition of n closed first order logic formulae. In most industrial plants the system designer has to deal with such complex equipment.

The maintenance experts use only two primitive objects: *faults* and *messages.* All the notions applied in diagnostic can be derived from these two objects.

The fault is a piece or an element of the industrial plant which is out of order, broken ot unsettled and leads to decrease the oprimal behaviour of the plant. A fault is unequivocally established by the visual verification of the defect (e.g., a broken element) or the verification by elementary measure instruments (voltmeter, oscilloscope, etc.).

Many messages are delivered by the central computer system or are observable on the industrial plant. They are effectively *observable sings:* control lamps, printed messages, curve deviations, and so on. When normally a message should be observed (e.g. a printed message), the absence of this message can be considered as a message too. The message are modeled by first order predicates and the validity of a message is interpreted as the truth value of the corrensponding predicate. A fault is valid if the logical atom attached to the fault is true. Messages and faults are expressed in C-Prolog[18] syntax. Some messages may induce a new message and this induction is denoted by:

$$m:-m_1,m_2,\ldots m_n$$

that should be read *"the message m is valid upon reception of the messages* $m_1,m_2,\ldots m_n$*".* A fault p depends on a list of messages and this dependency is denoted by:

$$p:-m_1,m_2,\ldots,m_n$$

that should be read: *"the fault p is valid upon reception of the messages* $m_1,m_2,\ldots,m_n$*".*

5.2. The Concepts

Diagnostic reasoning needs more elaborated notions than the two objects introduced. These notions can be derived by logical formulae from the objects[19,20].

5.2.1. The Symptom

The more interesting messages for troubleshooting are called *symptoms.* As the symptoms denote a message subset, they are defined by first order logic predicates.

Example: The predicate: print_message ("out of service analyzer n^o", V)

expresses the message "analyzer number V is out of service". But first order logic does not allow to define arithmetic expressions in a functional form. If the designer wants to measure an electric current through an analyzer, he cannot write:

$$current\ (analyzer)>20$$

because "current" is a predicate and not a function. Therefore we introduce two predicates, **val** and **valms,** to express a relation between a physical measure ans its value. Val is appropriate when the Prolog-variable is instantiated by the calculus of an arithmetic expression and **valms,** when a measure value is communicated to the system:

```
val (physical_variable, V), V is arithmetic_expression
valms (physical_variable, V)
```

Example: When the analyzer current value is 30 mA, the predicate

$$\text{valms (current (analyzer), 30)}$$

is added to the knowledge base. The comparison of this current to the reference value 20 mA is expressed by:

$$\text{valms (current (analyzer), X), X>20}$$

For some variables the last measured value is less important than the difference between the actual and the preceding value. For each variable, the system retains the actual measure (V_n) and the previous value (V_{n-1}). These two successive values are expressed by the predicate Δ as follows:

$$\Delta(a,D):-val(a,V_n),val(a,V_{n-1}),D \text{ is } V_n-V_{n-1}$$

Examples:

$\Delta(o_2,V),V<0$: oxygene drift is negative

$\Delta(rate(h_2),V),V>0$: furnace gas hydrogene rate is increasing

Other predicates may be defined, according to special needs. In this paper we shall use three basic predicates to represent the main gases involved in cast iron production:

$$\text{gas\_f(co), gas\_f(co}_2\text{), gas\_f(h}_2\text{)}.$$

5.2.2. Symptom Sets and Structures

If in diagnosis several symptoms are frequently used in combination, then the experts prefer to collect them in sets. These sets are defined by a list of predicates delimited by parantheses.

Example:

$$\text{(print\_message ("FG analyzer O}_2\text{ out of service"),}$$
$$\text{valms ((gas\_f(o}_2\text{),V1),V1>800)}$$

This symptom set is valid if the above message is printed, and the O_2 value is higher than 800. The main advantage of these definitions is the facility to define observed symptom collections.

Example:

1) **IF** message "FG analyzer O_2 out of service"

   AND O_2 furnace gas analysis > 800

2) **IF** message "AIR INLET"
   AND message "FG analyzer O_2 out of service"

   AND O_2 furnace gas analysis > 800

These symptoms are defined by:

```
        1-anal,
          2-(print_message ("FG analyzer O_2 out of service"),
          valms((gas_f(o_2),V_1),V_1>800);

        1-print_message ("AIR INLET"),
        1-anal;
```

5.2.3. The Context

A context is a set of symptoms related to a given state of the plant (e.g.: normal furnace behaviour after a slow running pace, furnace gas analysers invalid, etc.) or to independent events (e.g. rain, frost) which define the general conditions absolutely necessary to validate the specific fault symptoms. A context is a symptom structure shared by several faults.

The following expert rules:

    1) **IF** selection control lamps of the central computer are blinking
       AND the watch-dog is up
       THEN the central computer is in service.

    2) **IF** the central computer is in service
       AND there is a message ("COUPLER ABNORMALITY")
       THEN there is a coupler defect.

    3) **IF** "&" are printed
       THEN printer is saturated

are translated as:

```
    context 1 - in_service (central computer),
              2-(blink(selection_control_lamps),
              up(control_watchdog));

    context 1 - in_service (coupler),
              2 - in_service (printer),
                3 - in_service (central computer)
                3-in_service (printer), not print_message ("&"),
              2-not print_message ("COUPLER ABNORMALITY");
```

5.2.4. Fault Indices

Large industrial plants are usually designed with automata and industrial process computers. Messages are displayed by the central computer when determined faults may occur. These messages are not always the consequence of a defect and a fault index may occur when the fault supposed did not occur. This non-correspondence between a fault and a message has two major causes. First, the equipment was designed by a staff member of different disciplines and no expert has a whole knowledge of all the components. Secondly, frequent equipment changes (e.g. automata) may alter some message significances.

For example, in the central control room the message "air inler" is printed either when there is an air inlet, or when a three-way electonic gate is out of service.

From now on we use only the notion of *fault index,* because the object "*fault*" is seldom used by the maintenance experts. In the control room the experts see fault indices and reason only from that notion in their diagnostic process. Somt fault indices in our application are:

> — fault (three_way_electronic-gate)
>
> — fault (inlet (air))
>
> — fault (inlet (water))
>
> — fault (sensor (V)).

5.2.5. The Index Rule

The index rule[21,22] specifies the induction from the *context* c and the *symptoms* s_i to the *fault index* p:

$$p:-c, \ s_i$$

If several indices have the same symptoms, this means that the designer lacks information to discriminate them.

Index rule examples:

1) **IF** the central computer system is in service
   AND message "sensor n° xxx out of service"
   THEN defect on sensor n° xxxx or on the fox.

2) **IF** defect on sensor n xxxx or on the fox
   THEN measure voltage above the fox.

3) **IF** automaton IMM10 voltage=200∗ measured value −2
   THEN defect on sensor xxxx.

4) **IF** not automaton IMM10 voltage= 200∗ measured value −2
   THEN defect on the fox.

5) **IF** the central processing computer is in service
   AND furnace gas analyzers SMC50 are in service
   THEN furnace gas processing system is in service.

6) **IF** furnace gas processing system is in service
   AND mtssage "F.G. analyzer O_2 out of service"
   AND fg_analyzer value>800
   AND not message "F.G. AIR INLET"
   THEN defect on the fox.

Here, **F.G.** or **fg** are abbreviations for *furnace gas*. The preceding rules are translated as follows:

> **Context** 1 – in_service (central_computer_system);
> 2 – in_service (coupler),
> 2 – not print_message ("fault reading IMM10: see the report")
> **symptom** 1 – print_message ("sensor out of service n°:", V),
> 1 – mt,
> 2 – (valms (voltage (IMM10_entry), T1),
> valms (voltage (fox_entry), T2),
> T1=200∗T2−2)
> **index** fault (sensor (V));

```
      context in_service (central_computer_system);
      sympt 1 - print_message ("sensor out of service n°", V),
           1 - not mt,
      index fault (fox);


      context 1 - in_service (furnace_gas_processing_system);
                 2 - in_service (central_computer_system),
                 2 - up (smc_50_watch_dog);
      sympt 1 - o₂_analysis,
                 2 - (print_message ("F.G. analyzer out of service"),
                 valms (gas_f (o₂), V), V>800);
           1 - not print_message ("F.G. AIR INLET");
      index fault (fox);
```

The "central_computer_system" expression holds for the central computer, the coupler and the IMM10 automaton. "Furnace_gas_processing_system" denotes all the computer and automatic devices including the furnace gas analyzers. Through these examples, we see that human experts do not express all the knowledge needed in the form of inference rules; an advantage of a formal approach is to oblige the designer to completely specify the expertise.

5.2.6. Correlated Fault Indices

Maintenance experts often speak of parallel faults or induced faults to qualify events occurring at the same moment of resulting from the same original fault (23,24). It is sometimes very difficult to establish the chronological order of the events and to locate precisely the fault set which induces other defects. On the other hand, experts can express quite easily fault indices sharing the same symptoms. We call them *correlated fault indices*. They may hide one or several symptoms related to original faults, induce new symptoms not related to the original faults (e.g. a local power supply fault and some other defect may induce the message: sensor V out of service) or modify some symptoms. If fault indices are not correlated, they are called *independent*.

When a correlation between p_1 and p_2 induces a fault p_c, the symptom changes are denoted by:

$-S_j \leftarrow S_i$: symptom change

$- \leftarrow S_k$: S_k is no more valid

$-S_\ell$: a new symptom appears.

We now give an example of the central computer diagnosis during a restart procedure. The analyzer correlated fault is *"memory parity fault"* and *"software fault"*.

1) **IF** restart procedure n° 1
   AND MSRV2
   AND N=2
   AND HVO
   THEN memory parity fault

2) **IF** restart proceduree n° 1
   AND MSRV2
   AND N=A
   AND HVO
   AND the answer to "is there the repeated printed message disk out of service?" is "no".
   THEN software fault.

These expert rules may be translated into:

```
context restart_procedure (1);
sympt 1 - mv,
       2 - (msrv2, hv0),
       1 - (valms (n,N), N=2);
index fault (memory_parity);

context restart procedure (1);
sympt 1 - mv,
       1 - (valms (n,N), N="A"),
       1 - not_dk_printing
       2 - answer ("is there a repeated printed message disk
           out of service?"),
index fault (software);

context 1 - restart procedure (1);
original-indices fault (memory_parity), fault (software);
correlated-sympt((valms (n,N),←N="A", ←N=2);
corr-index (software_parity).
```

The "*answer*" predicate is valid if the reply to the question is confirmed.

5.2.7. The History

To detect more difficult faults, the experts need some successive variable values[25]. These values are called the *history* and are defined by an ordered set of valid symptoms stored on a direct access file. The history is denoted by the predicate:

```
hist (name, measure_date_measure_number, value)
```

name	: name of the analyzed variable
measure_date	: states the date and enables synchronization of different variable values
measure_number	: the last measure corresponds to the first record in the direct access file
value	: the measured variable value.

The predicate "*hist*" is valid if all the arguments are instantiated.

Example: For furnace gas analysis a control takes place each four hours. A drift of the gas values induces a calibration followed within three minutes by a new control. The timing is therefore expressed in minutes.

D_p	D_o	D_1
X	X <240 min>	X
E_p preceding calibration	C_o: control (normal)	C_{11}: triple drift control (control+3 drifts messages)
		E_1 : calibration
		C_{12}: control

The previous calibration E_p may occur more than 4 hours before the drift measure. The expert rules to diagnose the three-way-electro-gate state are:

1) **IF** furnace gas processing system is running
   AND message "drift CO"
   AND message "drift CO_2"
   AND message "drift H_2"
   AND CO, CO_2, H_2 values are different from those of the previous and following controls
   AND CO, CO_2, H_2 values from previous and following controls are identical
   THEN the values of the three drift control are uncertain.

2) **IF** the values of the three drift control are uncertain
   AND the durations of the three drift controls and calibrations are equal to the durations of preceding controls and calibrations
   THEN there is a fault on three way electro-gate.

3) **IF** message "F.G. AIR INLET"
   THEN furnace gas analyzer measures are invalid.

4) **IF** message "$COCO_2H_2$ analyzers out of service"
   THEN furnace gas analyzers are invalid.

To express these rules, the histories of the following three variables are needed:

    — control (contr), control_duration (d_contr)
    — calibration_duration (d_calib).

    Context 1 - in_service (furnace_gas_processing_system),
            2 - in_service (central_computer),
            2 - in_service (furnace_gas_analysers),
                3 - not (print_message ("F.G. AIR INLET"), print_message ("F.G. analyzers out of service", V));

    sympt 1 - ctr,
            2 - (print_message ("drift CO"), print_message ("drift CO_2"), print_message ("drift H_2"),
            2 - hist (contr (gas_f(G)),-, 3,C_o),
            hist (d_contr (gas_f(G)),-,3,DC_o),
            hist (contr (gas_f(G)), D_{11},2,C_{11}),
            hist (d_contr(gas_f(G)), D_{11},2,DC_{11}),
            hist (contr (gas_f(G)),D_{12},1,C_{12}),
            $D_{12} < D_{11} + 5, C_{11} \neq C_o, C_{11} \neq C_{12}, C_o = C_{12}$,
            hist (d_calib(gas_f(G)),-,2,$DCAL_p$),
            hist (d_calib (gas_f(G)),D_{11},1,$DCAL_1$),
            $DCAL_1 = DCAL_p, DC_{11} = DC_o$);
    index fault (three_way_electro-gate).

            $CAL_1 = CAL_p, DCAL_1 = DCAL_p, DC_{11} = DC_o$);

    The semi-colons in context 1.3 formalize a logical OR. The variable G take the successive value co, co_2, h_2.

5.2.8. The Fault Index Classes

Complex industrial plants are divided into functional components.
For further analysis the experts are led to collect the component indices
in a fault index class, because these symptoms are very often of the same
type. However the interrelations between the different symptoms change.
The *property set*[26] of an index class acts as the *context* and *symptom
structure* common to all the fault indices of this class. This set may be
empty.

We illustrate this concept through the example defined in 5.2.6. The
analyzed faults belong to the functional component "central_computer". If
they are the only analyzed faults of this component, the corresponding
class properties are:
          for the context: restart_procedure (1);
          symptoms; (msrv2, hv0).

5.2.9. Control Rules

At the control level, the designer includes the *strategy rules*
which are triggered when no fault index has been diagnosed definitely.
This particularly when one or several important symptoms have not been
analyzed by the object level.

The control level input hypotheses are fault indices worked out and
collected into index classes by the object level. The control rule conclusions are fault indices or context changes. These indices may act as
new hypotheses for further diagnosis.

Examples:

Let C be the set (T_f, C_f, S_i) where
T_f is cast temperature

C_f is carbon cast

S_i is silicon cast

and X, Y two variables among the three, Z being the remaining variable.

C(X,Y,Z) is defined by

$C(T_f,C_f,S_i) . C(T_f,S_i,C_f) . C(C_f,S_i,T_f) .$

$C_1(X,Y,Z):-C(X,Y,Z), X=Y.$

The following expert rules belong to the general fault index class, because the hypotheses involve two functional components: *blast furnace
process* and *instrumentation*.

    1) **Context** 1 - normal_state (blast_furnace),
          2 - normal_sate (ore_and_coke_loading),
          2 - normal_state (temperature (air)),
          2 - in_service (analyzers);
   hyp index_class (f_gas_analysis), fault (temperature (hearth));
   tests $\Delta(w_u,V_1)$, $V_1 \neq 0$, val(w_u,V_2), $V_2 > 520$;
   conclusion (fault (cooling (hearth)); fault (analyzer));

    2) **context** 1 - normal_state (blast_furnace);
   hyp index_class (f_gas_analysis), fault (temperature (hearth));
   tests $\Delta(w_u,V_1)$, $V_1 \neq 0$, val(w_u,V_2), $V_2 < 420$;

conclusions (fault (heating (hearth)); fault (analyzer));

3) context 1 - normal_state (blast_furnace);
   hyp index_class (f_gas_analysis), fault (temperature (hearth);
   tests $\Delta(w_u,V1)$, $V1 \neq 0$, Cl(x,y,z), X<0, Y<0, Z=<0;
   conclusions fault (cooling (hearth)).

4) context 1 - normal_state (blast_furnace)
   hyp index_class (f_gas analysis), fault (temperature (hearth));
   tests $\Delta(w_u,V1)$, $V1 \neq 0$, Cl(X,Y,Z), X>0, Y>0, Z>=0;
   conclusions fault (heating (hearth));

5) context 1 - normal_state (blast_furnace);
   hyp index_class (f_gas_analysis), fault (temperature (hearth));
   tests $\Delta(h_2,V_2)$, $V_2 > 0$,
         Δ(rate_per_thousand (coke), V_3), $V_3 > 0$,
         $\Delta(w_u,V_4)$, $V_4 < 0$, Δ(rate (co), V_5), $V_5 = 0$);
   conclusions context change out_of_service (analyzers);

6) context 1 - normal_state (blast_furnace);
   hyp cooling (hearth);
   test Δ(rate (co), V_2), $V_2 < 0$,
        $\Delta(w_u,V_3)$, $V_3 > 0$, $\Delta(h_2,V_4)$, $V_4 < 0$);
   conclusions fault (cooling (hearth));

7) context 1 - normal_state (blast_furnace);
   hyp fault (analyzer);
   test 1 - ($\Delta(w_u,V_1)$, $V_1 \neq 0$, Δ(rate (co), V_2), $V_2 \neq 0$,
        $\Delta(h_2,V_3)$, $V_3 \neq 0$),
        (hist (contr (o_2) ,_,1,V_4), hist(contr (o_2),_,2,V_5),
              hist (contr (o_2),_,3,V_6), hist (contr (o_2),→,
              4,V_7),
              $V_4 \neq 0$, $V_5 \neq 0$, $V_6 \neq 0$, $V_7 \neq 0$);
   conclusions context change out_of_service (analyzers);

8) context out_of_service (analyzers);
   hyp true;
   test true;
   conclusion (fault (dilution (f_gas (G))); fault (sampling-
              lines));

Explanations:

Limitation of the inversigation field:
In examples 1 and 2, different hypotheses must be refined. If Wu>520,
the field is reduced to two hypotheses: *hearth cooling* or *analyzer fault*.
The rules 3 and 4 guide the search to a single index.

Fault confirmation:
The maintenance technicians often need the confirmation of a fault,
before they undertake works that require long and expensive resources. If
rule 6 is valid, then the system displays this fault and continuous the
diagnosis with the remaining index classes.

Context change:

Given a certain number of hypotheses and after the verification of the symptoms (rules 5 and 7) the designer deduces a context change, for instance, the blast-furnace gas analyzers are out of service.

Change of the investigation field:

At the beginning the designer considered that the blast furnace context was normal. Because of the context change, the investigation field changed: the fault must be searched on the instrumentation level or in its environments (rule 8).

5.2.10. Summary

From the two objects, *fault* and *message*, we deduced all the concepts commonly used for fault diagnosis. Figure 1 displays the introduced notions.

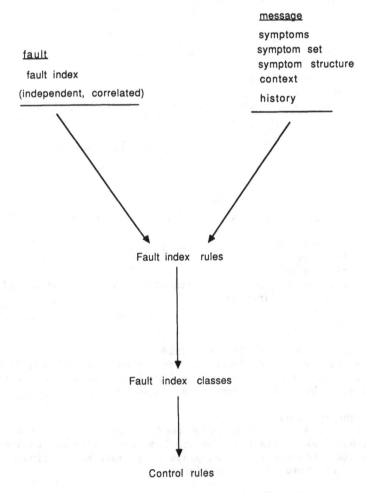

Fig. 1. Object and concept definitions.

6. SYSTEM ARCHITECTURE

The knowledge acquisition method previously defined can be used with any available expert system shell[26] provided that it offers logical deduction, arithmetic expression evaluation and propositional or first order logic, depending on the application complexity. The implementation will be more or less easy according to the existence of frames or other structure facilities.

Currently we develop a C-Prolog implemented diagnostic system generator, which will simplify the use of the concepts introduced in this paper. Figure 2 indicates the system's main functional architecture.

6.1. The Object Level

Starting with a context and symptom as input, the execution of the object level is as follows:

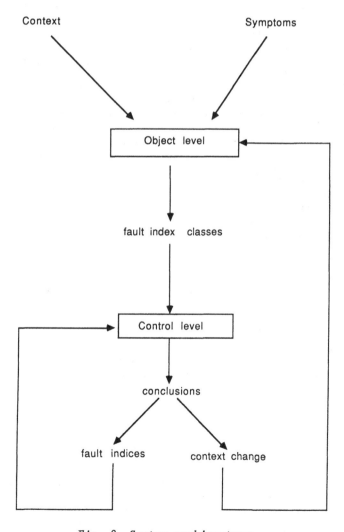

Fig. 2. System archirecture.

— search for the fault indices with at least one valid symptom

— by backtracking the system verifies, if all the symptoms of the selected fault indices have been introduced in the system. This step allows the maintenance staff to adjoin the symptoms they had neglected.

Even if some symptoms induce new symptoms, the execution of the object level is deterministic (no loop evaluation). If the system converges to a unique fault index, then this fault is considered as the effective one. Otherwise the indices are ordered in their functional component index class and the control level is activated.

6.2. The Control Level

At this level the designer introduces more complex knowledge involving separate functional components or whole plant related expertises. The control level execution is summerized by figure 3.

A control step is the application of the control rule. The execution of the control level is defined by the following algorithm:

— for each fault index class C,
  — choose the contex Ctx most indices of the class depend
    — for each fault index F in C that depends on Ctx
      — for each control rule R that has F as hypothesis,
        — if the remaining hypotheses of R are also valid, then draw the conclusions,
        — the conclusions are new hypotheses for the next control step,
        — the control steps continue until one of the halting conditions given below is satisfied,
      — end control rule R,
    — end fault index F,
  — end context Ctx,
— end class C.

For each class, the deduction stops if:

— a hypothesis has been confirmed as a fault index by a control rule

— the conclusions are the same as the hypotheses. The necessary knowledge to go further in indentifying the indices is missing.

The deduction continues if the number of hypotheses has been limited without converging to one single fault index. If during the deduction step a context change occurs, then the current context Cc is not valid. New indices are proposed and the system backtracks:

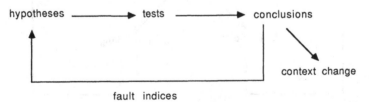

Fig. 3. Control level execution.

— the remaining rules are left out

— the next fault indices are left out

— all the fault indices valid under context Cc are retrieved.

The new valid Cc is the context proposed by the context change. The system returns to object level and verifies if the symptoms of the new indices are valid.

7. CONCLUSION

This paper analyzes a method founded on first order logic to conceive large industrial environment diagnostic systems.

Before the development of an operational system, the semantics of the concepts involved have to be defined. In our application field two objects, *fault* and *message*, are sufficient to deduce by logic the concepts used by the maintenance staff. It seems quite difficult to give a complete description of the structure and behaviour of large industrial plants only by logical formulae. Therefore we introduced the two categories of diagnostic approaches, the *heuristic* and the *fundamental*.

The peripheral equipment (process computers, automata, etc.) is analyzed by the heuristic approach. The plant behaviour is simulated by a mathematical model. This model allows a more fundamental approach for the fault detection. A control level takes this model into account. The system compares the observed values with those calculated by the mathematical model and states the divergencies with regard to the theoretical plant behaviour. If wrong or partial data is introduced then the possible dead ends are solved by a context change.

Our method defines a formal system which allows us introducing integrity constraints and rule consistency founded on a sound basis. Our future objectives are the development of these two complements which are necessary for permanent operational systems, and the introduction of *learning by analogy* facilities to detect new faults.

8. ACKNOWLEDGEMENT

I would like to express my gratitude to Monique Grandbastien for her time and energy during the discussion of this work, to M. Courrier at SOLLAC for the blast furnace expertise and to Pierre Lescanne for his comments on the chapter.

REFERENCES

1. F. Hayes-Roth, D. A. Waterman, D. B. Lenat, Building Expert-Systems, *Addison-Wesley* (1983).

2. J. S. Aikins, Prototypical Knowledge for Expert Systems, *Artificial Intelligence 20*, pp. 163-210 (1983).

3. M. Genesereth, The Use of Design Descriptions in Automated Diagnosis, *Artificial Intelligence 24*, pp. 411-436 (1984).

4. B. G. Buchanan, E. H. Shortliffe, (Eds.), Rule-based Expert Systems: The MYCIN Experiments of the Standford Heuristic Programming Project, *Addison-Wesley*, (1984).

5. J. Breuker, B. Wielinga, Use of models in the Interpretation of Verbal Data, in: Knowledge Elicitation for Expert Systems: a Practical Handbook, (Kidd A. ed.), *Plenum Press*, New York, (forthcoming).

6. B. Wielinga, J. Breuker, Models of Expertise, *ECAI 86* Brighton (U.K.), (1986).

7. J. G. Ganascia, Reasoning and Result in Expert-Systems: Main Differences between Diagnostic Systems and Problem-Solvers, *ECAI 1984*, pp. 31-40.

8. R. Reiter, Foundations for Knowledge-Based Systems, in: Information Processing, (Kugler H. ed.), *Elsevier Science Publishers, North-Holland* (1986).

9. B. Chandrasekaran, S. Mittal, On Deep Versus Compiled Approaches to Diagnostic Problem Solving, *International Journal of Man Machine Studies 19*, pp. 425-436 (1983).

10. J. De Kleer, J. Brown, A Qualitative Physics Based on Confluences, *Artificial Intelligence 24*, pp. 7-83 (1984).

11. K. Forbus, Qualitative Process Theory, *Artificial Intelligence 24*, pp. 85-168 (1984).

12. R. Davis, Diagnostic Reasoning Based on Structure and Behaviour, *Artificial Intelligence 24*, pp. 347-410 (1984).

13. R. Reiter, A theory of Diagnosis from first Principles, *Artificial Intelligence 32*, pp. 57-95, (1987).

14. M. R. Herbert, G. H. Williams, An initial Evaluation of the Detection and diagnosis of Power Plant Faults Using a Deep Knowledge Representation of Physical Behaviour, *Expert System 4*, (2), pp. 90-98 (1987).

15. F. Jakob, P. Suslenschi, D. Vernet, Extase: An Expert System for Alarm Processing in Process Control, *ECAI 86*, Brighton (U.K.) (1986).

16. M. Grandbastien, J. Maroldt, Rowards an Expert System for Troubleshooting Diagnosis in Large Industrial Environnements in: Applications of Artificial Intelligence in Engineering, (D. Sriram, R. Adey eds.), *Springer Verlag* (1986).

17. J. Maroldt, Proposition d'une méthode et d'outils pour la spécification de systèmes orientés diagnostic, 2. *Colloque International d'Intelligence Artificial de Marseille*, (1986).

18. W. F. Clocksin, C. S. Mellish, Programming in Prolog, *Spinger-Verlag*, (1984).

19. J. Mccarthy, Applications of Circumscription to formalizing Common-Sense Knowledge, *Artificial Intelligence 28*, pp. 89-116, (1986).

20. H. Gallaire, Pol: Logic, Objects and Semantic Models, *AAAI 86 Proceedings*, Philadelphia, (1986).

21. W. J. Clancey, The Epistemology of Rule-based Systems, *Artificial Intelligence 20*, pp. 215-251.

22. J. L. Lauriere, Un langage déclaratif: SNARK, *Techniques et Sciences Informatique*, vol. 5 (3), (1984).

23. G. Kahn, On when Diagnostic Systems want to do without Causal Knowledge, *ECAI 84*, pp. 21-30, (1984).

24. J. De Kleer, B. Williams, Diagnosing Multiple Faults, *Artificial Intelligence 32*, pp. 97-130 (1987).

25. J. Allen, Maintaining Knowledge about Temporal Intervals, *CACM 26*, pp. 832-843 , (1983).

26. P. Langley, B. Nordhausen, A Framework for Empirical Discovery, *Imal 86*, Les Arcs (Savoie), (July, 1986).

27. D. A. Waterman, A *Guide to Expert Systems*, Addison-Wesley, (1986).

6

CHAPTER 6

EXPERT SYSTEM BASED ON MULTI-VIEW/MULTI-LEVEL MODEL APPROACH

FOR TEST PATTERN GENERATION

C. Oussalah*, J. F. Santucci*, N. Giambiasi* and P. Roux**

* L.E.R.I., Parc d'activités Scientifiques et
   Techniques, 74 chemin bas du mas de Boudan
   30000 Nimes, France

** CIMSA-SINTRA, 10 Avenue de l'Europe
   78140 Vélizy, France

1. INTRODUCTION

A fundamental requirement in any CAD system is the possibility of defining a useful model to represent the behavior of real world systems {1}.
The resolution of complex problems in the technical domains requires the
use of hierarchical and multi-view models {2,3,4}. At each level of abstraction different sub-problems have to be solved; therefore the resolution technique consists essentialy of a problem reduction approach. In
the first section of our chapter we deal with a general structure for
model, which can be summarized as consisting of different model views,
each view being represented at a number of distinct levels of abstraction.
With this type of structure, complex applications may be spread over different levels of abstraction and different model views. We call this kind
of design the *multi-view/multi-level modelling* approach.

In general, a system model can be composed through a number of distinct views, each of which represents a particular aspect of the system;
in the system modelling we have used three types of models, namely a *behavioral model*, a *structural model* and a *physical model*. We give particular attention in models of *network type* including an explicit description
of different entities and links that appear in this model. Two types of
links between representations can be distinguisted:

— *Vertical Links*: allowing information to be transferred between levels
  of abstraction in a given model view.

— *Horizontal Links*: by which information can be transferred between
  representations in different model views.

Another important concept in our framework is the elaboration of
model libraries in which all the models developed or used are saved along
with the modelling of a more complex object. This model may be seen as a
real knowledge base of the application designers.

The final stage in the definition of our general model structure is
to use the graph-directed formalism for describing each representation of
the model.

In the second section we present an expert system for test pattern generation based on the general structure mentioned above {5}.

Algorithmic tools, like ATPG (*Automatic Test Pattern Generator*), which are perfectly adapted to problems for which they are designed, fail when they are applied to complex digital boards. Therefore, due to the complexity of an interconnected integrated circuit system, algorithm tools cannot be used on such sets of components. To overcome this complexity problem the only realistic approach is a hierarchical approach with problems reduction using A.I. techniques {6,7,8,9}. But this must be done in harmony with the classical tools available in this domain {10}.

Thus the problem has to be considered at different levels:

(1) Test Pattern Generation for sets of components whose size is acceptable for ATPG.
(2) Test Pattern Generation for digital boards.
(3) Test Pattern Generation for a computer (which is a digital board interconnection).

The generation at the first level is well known; it can be done by an ATPG. The problem is to solve the test pattern generation at the second and fird levels for which ATPG cannot be used.
We present in detail a test pattern generation system at the board level.

This problem is reduced into a set of three sub-problems:

(1) Elaboration of a board model {11}.
(2) Test pattern generation for sets of components of the board whose size is acceptable by an ATPG.
(3) Resolution of the controllability and observability problem.

The problem of test pattern generation for digital boards is solved by an expert system {12} which implements the first and third point and uses an ATPG in order to solve the second point.
The various concepts and components of our expert system are presented in this chapter.

2. HIERARCHICAL/MULTI-VIEW MODELS

A digital system can be observed from different points of view and represented by various designs of model; for instance, a computer can be represented by two models:

(a) a thermal model for the stydy of the heat and its dispersion
(b) a functional model which simulates the command and the functioning of the computer.

It is often necessary to modelize a system by different abstraction levels. Experts use this approach to tackle intricate problems. By defining a proper abstraction level for an application step, the semantic properties of the data to treat are limited by releagating non-relevant details at the inferior abstraction level. Models designed by this kind of structure are called *hierarchical models* or *multi-level* models. In other respects, the model of a real system is generally composed using a set of destinct views corresponding to its different aspects during the successive application steps. Models designed through this point of view are called *multi-view* models. Meanwhile, in several existing representation systems, there is no difference between the notion of a *model view* and the notion of a model abstraction level. These two notions are often overlapped and making them explicit allows one to clarify the concepts belonging to different model views.

2.1. Notion of Model Abstraction Levels and Views

2.1.1. Notion of Abstraction Level

The description of complex systems needs the construction of an abstraction hierarchy. The advantage of this point of view comes from the application of the partition rule to the whole model, i.e. the reduction of the problem to a succession of easy-tractable subtasks. Concerning the digital circuits, many works have led to appropriate levels in the abstraction hierarchy; for instance, some relevant levels in the circuits description have been established:

- — system level
- —— register transfer level
- — gate level
- — transistor level.

Other levels can eventually occur in a particular task. During the modelling process, the application designer sets up all the levels required by the model structure, and defines the tools that are appropriate for the process corresponding to each level. Moreover, data connected to each abstraction level can be represented in a variety of ways.

2.1.2. Concept of abstraction level

The abstraction level of a model includes:

(a) the data structure model
(b) the tools used by the application designer to specify the functioning of the model
(c) the structural detail level in the model.

2.1.3. Notion of model views

Among the most important selected views in the modelling process of a digital system, one can observe:

- — behavioural views
- — structural views
- — physical views.

These different views constitute the base for describing the different facets of a digital system.

(a) *Structural view*: By "structural" we mean the model topology. A structural model describes the inter-connections between various models (which can be structural or behavioural).
(b) *Behavioural view*: It specifies the behaviour and the input/output specifications of a "black box" like component; knowledge of the physical realization of the component is not essential to that kind of model.
(c) *Physical view*: It refers to the physical realization of a particular model.

2.1.4. Functional/behavioural interchangeability

In the litterature, the terms "functional" and "behavioural" are often interchageable. In order to clarify these two notions, we give their respective semantics through an example.

The functionality is a description level different from the behavioural one. For instance, the clock behaviour can be seen as a rotation about a point, but its function is to say to an observer what is the time.

In our study, it would be more relevant to deal with behavioural models than functional ones, considering that a model by itself has to be a functional model.

2.2. Methodology to Construct Hierarchical/Multi-View Models

Our aim is to define a methodology that makes easier the construction of hierarchical/multi-view models. The following steps are followed:

(1) Analysis and classification of the various representations of the model. The question is to first analyze the class of applications for which the model is constructed, and identify the particular representations which the model has to understand. These representations depend largely on the tools which will be used during the modelling process.

(2) Definition and elaboration of libraries concering the application classes. These libraries allow one to give behavioral and structural information concerning elementary components which constitute the model. In most cases the systems to be modelled are very intricate and the difficulty to estimate the whole diversity of knowledge is overcome by using various representation schemes more or less adapted to this or that application. In order to simplify the model description, it is essential to store data independent of context in appropriate libraries. In this way, the only data needed to be specified by the user during the description process of the model are "contextual".

Actually, three classes of functions related to a library can be defined:

(a) Functions for defining the library structure
(b) Functions for constructing the library, and
(c) Functions for using the library.

(3) Definition and representation of the different views of the model.
(4) Definition and representation of the different abstraction levels of a model view.
(5) Elaboration of a data transfer mechanism between the different views and abstraction levels of the model. The basic features of the multi-levels/multi-view model are:

(a) The data transfer mechanism between the different abstraction levels or different views of the model.
(b) The possibility to define some data transformation methods for the various kinds of data.

Previously, we have observed that the general structure of the model is based on the definition of a set of views and abstraction levels, and on their interconnections.
Two interaction classes between the different representation can be pointed out:

(a) *Vertical interactions*: They allow to transform some model data from a level to another adjacent level.

(b) *Horizontal interactions*: They allow to transform some model data from a view to another different view.

In these two cases, processed data are subdued to a transformation function. Users' classes occuring in the model elaboration process are associated to the above different steps.

2.3. Concept of Model External and Internal View

A model is identified by an *external view* (generally called *black box*) and an *internal view*. The first view refers to an item (box) which has inputs and outputs. The second gives an internal description of this box.

Thereby, a model external view will be represented by a *node* and its *inputs/outputs*. On the other hand, a model internal view will depend on the type of the selected model; if it is a network model, its internal view will be represented by a *graph* and its *inputs/outputs*, otherwise, it will be represented by the *function/variable* pair effecting the model behavioural description.

2.4. Interactions in a Model

The problem of the interdependence between models is an important point for the GPS. As a matter of fact, in the modelling process, as soon as an item of the real word is treated, we are confronted with some diversified inter-relations the discovery and control of which can constitute an important part in the resolution of the problem. These dependencies can be located in the abstraction levels defined in the model between two adjacent levels or between two levels belonging to distinct views. Therefore, a communication protocol between the various nodes of the model has to be created, including prospection and retraction mechanisms. Two interaction classes exist:

— vertical interaction classes

— horizontal interaction classes.

The vertical interaction classes generate the following kinds of links:

— connection links

— decomposition links

— expansion links

— compression links.

(a) *Connection links*: They refer to existing links between nodes of the same abstraction level.
(b) *Decomposition links*: They represent existing links between a model external and internal view, knowing that it is a network model.
(c) *Expansion links*: They allow data transformation from an abstraction level to another inferior adjacent level. They represent links created between some external views of two abstraction levels. Indeed, the application designer can use a different data representation in two adjacent abstraction levels. At first, he has to write a (vertical) translator defining the transformation.

A *translator* permits one to define how the same data and protocol are represented in the two different data representations. Figure 1 is a refinement of data representation using translators.

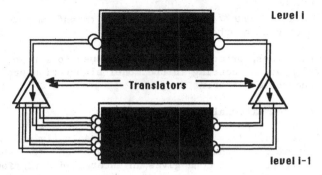

Fig.1. Refinement of data representation.

(d) *Compression links*: They represent opposite links compared with expansion links. So, the use of translators is essential.

The horizontal interactions class generates only one kind of links, namely that which executes data transfer between two abstraction levels belonging to two different model views. This kind of transfer also requires a (horizontal) translator writing.

In short, every model can be represented by an *external* and an *internal view*. The interactions that direct the communications between these two configurations are based on the links already explained. In order to consider all the relational descriptors occuring in a model, an interactions taxonomy operating on the under modelling is established. Then, we single out for distinction the following.

2.4.1. Interactions between a model external and interval view

Two cases can arise in this kind of interaction:

(a) The internal view is represented by a graph: The binding is made by an inter-relation between the model external view input/output and its internal view input/output, i.e. by its structural graph (see Fig. 2). This interaction is realized by the decomposition on the node, which represents the model external view, into a graph representing the model internal view.

(b) The internal view cannot be represented by a graph: The binding is made by a connection between the model external view input/

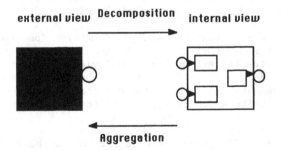

Fig.2. External/internal view of a node.

output and its internal view communication variables. This interaction is realized by a relation between a node, representing a function, and the modules realizing this function.

2.4.2. Interactions between two adjacent abstraction levels of a model $La_m(i)/La_m(i-1)$.

Notation:

$La_m(i)$: represents the abstraction level i of a model m.

Ev_M : represents the external view of the model M.

This kind of interaction is the source of the existing connections between the two external views $Ev_M(La_m(i))$ and $Ev_M(La_m(i-1))$. The link between these two adjacent levels always exists and allows the isomorphic transformation operating on the $Ev_M(La_m(i))$ and $Ev_M(La_m(i-1))$ input/output. This interaction is realized by either an expansion link during the passage from the level $La_m(i)$ to the direct inferior level $La_m(i-1)$ or by a compression link (opposite link) (see Fig.3).

2.4.3. Interactions between two abstraction levels of two views of a model $La_m(n_i,v_j)/La_m(n_i,v_m)$

Notation:

$La_m(n_i,v_j)$ represents the abstraction level n_i of the view v_j of the model m.

There is no method for automatic transformation between the various kinds of a model's two data views. Hence, a translation must be defined between the $Ev_M(La_m(n_i,v_j))$ and the $Ev_M(La_m(n_i,v_m))$ inputs/outputs. The data transfer between the two views of a model is generally *partial* or *local*, and consequently it doesn't dispose of general mechanisms permitting a complete exchange between the two views.

3. MODEL REPRESENTATION

In the general model representation system developed, models are represented in the form of graphs. A model can be described using *nodes* to represent the elements by which it is composed, and *arcs* to show interrelationships between them. Hierarchical models can be represented by using a set of *node/sub-graph* pairs. The sub-graph of a node represent an element as seen at a lower level of abstraction. So an extension of the graph formalism was chosen to describe each representation of the model. The graph structure is composed of:

— Nodes,
— connections,
— Connectors,
— Ports.

Node: A node is a *unit of description* of a basic element or a *function* belonging to a given level of abstraction of a model. It can be described either by its corresponding sub-graph or by an instantiation of a generic model assigning some of its attribute values.
Two types of nodes are identified:

— Context-Free-Node (C-F-N)
— Note-In-Context (N-I-C)

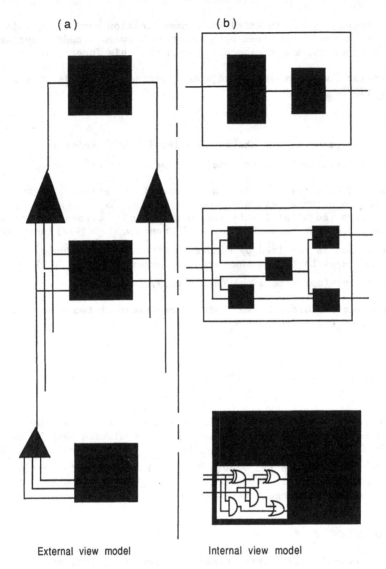

(a) (b)

External view model   Internal view model

Fig.3. Node/sub-graph correspondence: (a) + (b).

Context-Free-Node: It describes an elementary component in terms of information that is independent of any eventual use of the node.

Node-In-Context: It represents an elementary component in the context of a given model. It would describe, for example, the interconnections with other nodes in the graph.

Connection: A connection is an existing link between two nodes belonging to the same level of abstraction.

Connector: A connector is associated to both graphs and nodes and is composed of *ports*. It constitutes a well defined interface between a graph or node and the surrounding environment.

Port: A port is an entity attached to a node's connector and is the only means by which data can enter or leave nodes.
Three types of ports are defined:

— Source port   : a port through which data, is transmiteed,
— Destination port   : a port to which data is sent,
— Bidirectional port: a port enabling data transmission in both directions.

Figure 4 illustrates the structure of a node.

4. GENERATION AT THE BOARD LEVEL

4.1. Introduction

As we have seen, the technology evolution does no more allow us to base the test pattern generation only on the ATPG utilisation. Nevertheless, these ATPGs remain necessary subject to the condition that the waist of the subdued circuits is such that they can be tested during a rational time. To solve this problem, we have been induced to consider the three following steps:

— a functional partitioning
— a test pattern local generation
— an organisation of these vectors.

At present, the partitioning and organisation steps are essentially based on experts knowledge, while the generation step for each subset can be performed by tools which are available and largely used in the industrial environment. The generation system for complex digital boards has been developed by automating the first and the third step through expert systems.

4.2. Functional Partitioning

The solution of the test sequences problem is based on integrated circuit modelling at abstraction levels that correspond to those used by the expert. In our case, we are working on "two-board models". The first model represents the board at the physical level (i.e. it is described in physical package interconnections framework. From this model which we call *"board structural model"*, a superior abstraction model is built. This model, called *"macro-topologic board model"*, is obtained by the physical package agglomeration based on the experts'knowledge in order to obtain a board model where the basic items are subsets of packages called *macro-components.*

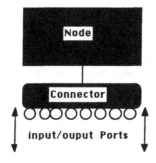

Fig.4. Structure of a node.

Before the functional partitioning step, the modelling and physical board description steps must be completed.
The functional partitioning step involves:

— creation from the physical board description of an internal model

— exploration of control and data path in order to determin some components to be likely agglomerated

— application of partition rules to confirm the components agglomeration

— macro-components construction.

4.2.1. The board structural description

The model elaborated so far is called *board structural model*. The user describes the board giving a list of the design issues and the number of packages as their inter-connections at their pins level and also at the board input/output level. At the package pins level, this data is provided by the components library. The "board pins" definition is automated by the analysis of the kind of package pins they are connected with.

4.2.2. Topological model construction

In this step, the system attempts to reproduce the human experts' reasoning. For the greater part, expert distinguishes two kinds of areas in the board: *data* and *controls*.

The *data area* is composed by components dealing essentially with data flows. The *control area* refers to components which create control flow supervising the data area. Consequently the aim at the first stage to obtain a board model that presents the different data and control flows.

From this model, called *internal model,* the functional model is constructed by the following three modules:

— strategic module
— reasoning module
— agglomeration module.

A. Internal model. The internal model is a data structure specific to the functional partitioning, and includes

— useful data generation from the data stored in libraries, i.e. general data concerning digital circuits (such as the potential bus notion) and data specific to the functional partitioning (such data defining how some packages can be agglomerated i.e. agglomeration rules described by the expert).

— the board description, provided by the expert, which has been translated as a graph (viz. the board description).

The board physical description is transformed to an internal model. This board internal representation allows one to optimize the treatment process by additioning virtual nodes (such as divergence nodes which permit one to make clear the component multiple outputs). The functional partitioning is based on this divergence concept.

B. Functional partitioning.

B.1. Strategy module. This module drives the strategy of research in the functional partitioning step. It proposes some components to the reasoning module which could be agglomerated. According to expert's approach, the strategy module scans at first the data area and then the control area.

The data area treatment: The research process is to watch over the data flows considering only one data flow at a time. The scanning is performed at first by a deep search from the board inputs to the outputs, progressing in a component layer successive way. Macro-components are gradually generated after the reasoning module application (which confirms if the selected component must be agglomerated) and after the agglomeration module application (which produces the macro-component).

The treatment is based on two concepts, i.e. the *flow splitting point* and the *flow full divergence*. The splitting point corresponds to a partial divergence of the flow; the full divergence corresponds to a complete divergence of the flow. These two notions are depicted in Fig. 5. If neither a splitting point nor a full divergence occurs, all the components through which the flow is going are agglomerated. When a full divergence occurs, all the flow branches are separately explorated. When a spitting point occurs, the components which will reconsitute the data flow (if they are allgomerated) are selected. Then, some partition expert rules are applied to decide if the agglomeration must be done.

The control area treatment: The major aim of the control area treatment is to do new agglomerations between control area components or to include to the macro-components, generated by the previous treatment, the control packages which control them. The treatment is executed in three steps corresponding to three stages of the flow running:

— From the board control inputs to the outputs, by an analysis identical to the data area treatment.

— From the board output flows not yet treated to the inputs.

— From control inputs of the macro-components generated during the data area treatment of the board inputs (for eventual extension).

B.2. Reasoning module: When the components to be eventually agglomerated have been selected by the strategy module, the partitioning rules refering to the packages are applied. They finally decide if the agglomeration must be done or not. An example of agglomeration rule is given in Fig. 6.

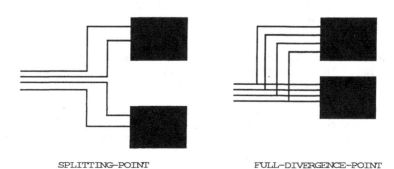

SPLITTING-POINT   FULL-DIVERGENCE-POINT

Fig. 5

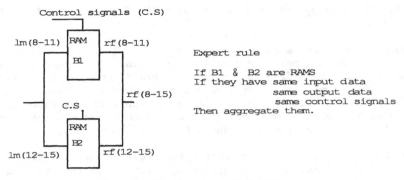

Fig. 6. Aggregation of 2 RAMs.

B.3. Agglomeration module: If the reasoning module decides to carry-out component agglomeration, the agglomeration module constructs the resulting macro-component.

4.3. Generation and integration of test vectors

After the construction of the model describing the board with macro-components and macro-connections, the problem is to generate test se-quences allowing one to obtain a sufficient fault coverage for the board. But once the macrotopological model is constructed, in which the basic items have a reasonable waist for an ATPG, the next task is to solve the observability and controllability problem and to determine the macro-component to subdue to the ATPG in order to obtain a sequence testing the whole board with a maximum fault coverage. To aid the solution of these two problems, the *test path* concept has been introduced. A test path is a data path permitting the propagation of test patterns from the board inputs to its outputs, described at the macro-component level. Once the test path is constructed, that permits to have a full fault coverage of the board, the macro-component to subdue to the ATPG must be selected. To this end, the system reproduces the reasoning of the expert who selects the macro-component by an empirical criterion. In our approach, this selection is done in two stages:

(a) Test path selection in the board: This selection is made by expert rules deducted from his/her experience. These rules take into consideration the test path complexity for selecting the path which will allow one to obtain the most complete fault coverage when the fault simu-lator is applied (this does not allow apply an ATPG to all the board macro-components).

(b) Choice of the macro-component in the selected test path during the first stage: This selection is again made by means of expert rules. The selected macro-component is called *target macro-component.*

Before submitting the target macro-component to an ATPG, the control-lability and observability problems must to solved. This determines the conditions of test pattern propagation through the board. The test pattern propagation uses expert rules for data transfer (backward and forward).

The controllability problem solution begins with the behavioural description into libraries of the physical packages which compose the board. This behaviour, described by the backward and forward propagation rules, is in fact a simplified behaviour which allow one to solve the problem without treating the general case or theoretical problems that never or rarely occur in real boards the experts have to treat.

Consequently, the problem is to determine conditions that allow one to bring a pattern, for instance the value A, to the target macro-component inputs. In other words, the problem is:

— to determine the pattern f(A) which is to be applied to the board inputs and reflects simple transformations (shift, implementation, etc.)

— to determine controllability patterns which consitute the conditions of propagation f(A).

To solve this problem, the behavioural description of integrated circuits is at first used to obtain the macro-component behaviour. After that propagation transformations and conditions are obtained, by a symbolic backward propagation through the path components in which the target macro-component belongs. Then, the target macro-component description is translated at the gate level and subdued to the ATPG for the test pattern generation, which is local to the target macro-component. The macro-component test pattern is propagated through the selected test path keeping the propagation transformations and conditions previously determined. At this stage, a fault simulator is used; on the one hand, testing sequences, and on the other, the path description at the gate level are provided to it. It gives us the list of tested faults. Then, our system takes over the organization of the whole board test pattern. For each component, a coverage rate is calculated taking into consideration already tested fault. Then, the results are analyzed in order to select either another target macro-component in the path (if the coverage rate of one of the path macro-components is not satisfying) or another test path of the same treatment (if the coverage rate is satisfying).

4.3.1. Test path construction

A. Test path notion

Definition: *A test path is a data path allowing one to propagate testing patterns from the board input to its output, described at the macro-component level.*

The testing paths are constructed so as to obtain a full coverage of the board. Moreover, they will help us to select the macro-component to subdue to an ATPG. Finally, the test path concept helps the treatment at the time of symbolic backward propagation. Our approach differs from classical ones which consist in choosing a macro-component in the board and making the choice of the path at the propagation time. To avoid a combinatorial explosion of possibilities, we select this test path from expert rules, and the macro-component from other expert rules.

B. Test path enumeration

Test paths are constructed from the analysis of the graph describing the board at the macro-component level. All possible paths from inputs to outputs are checked (we say "path" in the graph teminology). A path is watched over, beginning by its inputs:

— If there is no multiple input components on the whole path, this path is a test path;

— If a multiple input component is met, one must search from this macro-component the paths with the same successor and different predecessor subpaths. The set of these paths constitute a *test path*.
  Thus test path is a board subgraph.

In fact, test path determination is essentially based on the different kinds of nodes:

— *switch node (a kind of multiplexor)*: This node is fictitious during the board structural modelling in order to be able to model what is called a "tree-state". A tree-state corresponds to the tree-state bus notion. This notion is consequently modelled by a node called *switch-node*. This node has several inputs and only one output, to which a decoding function is associated that helps to remember what input is selected.

— *multiple input node*: For instance, an ALU; in this case, all the predecessor subpaths must be considered.

4.3.2. Path and macro-component selection

This selection is based on expert knowledge. As a matter of fact, this selection is made empirically by the expert. This empirical knowledge can be updated by the expert's experience. This is the reason why this step is realized by means of declarative concepts.

A. Test path selection

In order to simulate the expert knowledge, different selection criteria have been associated to the test paths. As we have seen previously, the aim is to select test paths permitting to detect a maximum number of faults during the fault simulation. The criterion definition is based on the single - or multi-input - macro-component notion, and on the marked and non marked macro-component concept. A single-input macro-component is a macro-component having only one data input (for instance, a buffer).
A multi-input macro-component is a macro-component having several data inputs (for instance, an ALU).
A market macro-component is a macro-component transforming data (for instance an ALU).
A non marked macro-component is a macro-component which doesn't transform data (for instance, the serial or parallel registers).
Marked macro-components are the most complex for testing according to the expert's experience. The selection criteria wiil have to take this aspect into account.
Since the aim is to select test paths allowing to detect a maximum number of faults during the fault simulation, the selection criterion depends on these two notions.

The various criteria are:

— The type of the path, which expresses the path complexity. It is represented by a calculated number from the type rules given by the expert. These rules take into consideration the kind of the macro-component (which can be marked or not, single or multi-input).

— the complete number of macro-components.

— the number of marked tested macro-components.

— the distance from the first marked and non tested macro-component to the board inputs.

— the complete number of non tested macro-components.

Afterwards, the problem is to determine how these different criteria are combined together to select the test path. To realize this, according to the expert's experience, some weighting values are assigned to the different path selection criteria.

B. Target macro-component selection

The macro-compoment to subdue to the ATPG, called target macro-component, is selected after the path selection with the help of expert rules which depend on the type of the path. Once the test path and consequently the target macro-component is selected, there remains to solve the controllability and observability problem. The solution of this problem is again based on expert knowledge. As a matter of fact, the problem wiil not be solved for all possible hypotheses; that would be very hard due to the complexity of the problem. The problem is substantially simpified by sime simple rules about the components established by experts, thus permitting to solve it for nearly all hypotheses. So, the problem is not solved in the general case, but nearly all the cases which the experts know, are solved using this indirect approach.

4.3.4. The controllability problem

4.3.A. Behavioural description

Rules that simplify the behaviour given by experts are stored in the component library. The description stored in the library must include not only structural data but also behavioural data. The behavioural description that permits the data bi-directional propagation through the component, is processed by rules which are dependent on components stored in the library. So, the behavioural description is represented both by a formal description language, which specifies the component structure, and by production rules that describe the component function.

Example of behavioural description: The behavioural is described by the user according to the behavioural description language. This language is in fact a sub-language of HDL. Its syntax has been defined according to the kind of behaviour the CIMSA experts use. This language has two distinct parts:

— a declaration part

— a behavioural part.

The declaration part corresponds to the definition of variables which can be declared as *data-in, data-out, cmd, bidata,* and *initial.* Simple variables can be defined for the wire representation (example: A,X), and sized variables for buses (example: B(1:4)). Using the key word *Alias* permits to redefine the bus or a part of it (example: I(1:2) *Alias* (A,X),C *Alias* b(2:3)).

The behavioural part, permits one to describe a simplified behaviour with the help of the orders: *"if, when, make".*

B. The symbolic propagation through the test path

The symbolic test pattern propagation forward and backward is based on the transfer concept, and therefore on the relation between the various modelling levels. Symbolic values are associated to the *arc* and the *gates* of the different manipulated objects. The backward symbolic propagation of the value is hierarchical. It is done in two steps:

— at the component level

— at the path level.

(a) At the component level: From the bahavioural description put by the human expert in the library, symbolic values to be taken by the component inputs and the corresponding controls are detected. An example of backward symbolic propagation through a macro-component is given in Fig. 7.

(b) At the path level: The aim is to determine transformations and controls to apply at the pattern inputs in order to obtain the desired values for the target macro-component inputs. So, the symbolic values allocated to the target macro-component must be propaged backward through the different nodes of the path.

For each node type in the model, there is a function, called *propagation function* of the node. This function provides, according to the symbolic values given to the output gates of the node, the symbolic values and controls to be given to the input gates of the node that correspond to the data inputs and controls permitting the propagation in the macro-component. During the simulation, *conflicts* can occur. A *conflict* is a situation where a value V2 occurs in a connection where a value V1≠V2 was given. It corresponds to two different values already produced by the exploration of the two limbs of a divergent point in the physical board. A conflict can be resolved when a memorizable item can be found in one of the limbs towards the divergence point (examples: a register, a memory, etc.). In this case, the two conflicting values create different test patterns. Therefore, during the fault simulation, test patterns are successively applied according to propagation conditions determined for each pattern.

Propagation rules are:

— *For the first pattern:* Controls to apply on every memorizable component to memorize the corresponding value.

— *For the second pattern:* Controls to apply on every component in order to propagate the test pattern towards to the macro-componet inputs.

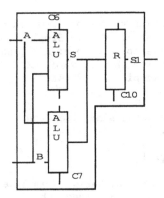

For propagating test patterns through this macrocomponent the expert leads the pattern p2=0 on bus B and the pattern p1 to propagate on bus A.
Then it acts on control signals of C6 and C7 in order to realise the addition p1+p2=p1.
Finally, the expert acts on control signals to propagate the pattern through C10.
Some problems may occur due to the wiring of control signals or input data. For instance ALU can only p2-p1. In this special case, the expert decides to lead the pattern to propagate on bus B and the pattern 0 on bus A.

Fig. 7 Backward propagation through a macrocomponent.

In the hypothesis where there are no memorizable items other behavioural rules must be tested to resolve the conflict. However, this backtracking process is not always a solution. If a conflict cannot be resolved, the user must propose a solution by himself.

4.3.4. Call to an ATPG and a fault simulator

The macro-component is transalated at the gate level and fed to an ATPG. The ATPG provides the test pattern that must be applied to the macro-component inputs. Then, the values to be applied to the board inputs are deducted. Thus, the test pattern description and the pattern previously determined are given to the fault simulator. The fault simulator lists for each component of the path, the faults tested by the test sequence.

4.3.5. Analysis of the fault simulation results

The analysis of the fault simulation results is essentially based on the expert's experience. At the analysis time, the expert does not only know the test path and the results of the simulation, but also has a complete view of the board. He knows, in particular, what happened on the other paths during the different calls to the fault simulator. To model this knowledge, the path state concept is associated ot the test path. This concept allows one to know at every time the treatment which has been done on the path. To automate the analysis phase, the transition from a path state to another state is modelled by a state graph. Links between different nodes of the graph are established by expert rules.

A. Checking the different states of the path

At the beginning, the board paths are in the *"Initial"* State (I). The initial state corresponds to the situation where no operation has been done on the path; i.e. no component of the path has been declared.

The *"Tested-Macro-component" State* (TM) corresponds to the situation where some macro-components of the path have already been tested on another path during the fault simulator application. When a path is selected, it evolves in the *"Current" State* (C). It corresponds to the situation where a path is being treated; the test path description is translated at the gate level and subdued to the fault simulator.

The *Terminal State* (T) represents the situation where all the components of the board have been tested during the fault simulation of the current path and on other paths having the same macro-component.

The *"backward-problem"* state (BACKWARD-PB) represents the situation where one of the macro-components, which lies before the target macro-component position on the path, has not a sufficient coverage rate.

The *"forward-problem"* state (FORWARD-PB) represents the situation where one of the macro-components, which lies after the target macro-component position on the path, has not a sufficient coverage rate.

The *BACKWARD-C* state represents the situation where a path has been reselected after a problem that occurs before the target macro-component position.

The *"backward-forward-problem"* state (BACKWARD-FORWARD-PB) represents the situation where a *backward* or *forward* macro-component has not a sufficient coverage rate.

B. Definition of the transition rules

R1: All the path states are initialized by the I state.

R2: A path in the I state, having one or more macro-components tested on another path during the fault simulation, turns into the TM state.

R3: When a path is selected, it turns into the C state. To be selected, a path must be in the I or TM state.

R4: If a path is in the I state and the results of the fault simulator lead to a backward problem compared with the target macro-component, the path turns into the BACKWARD-PB state.

R5: If a path is in the I state and the results of the fault simulator lead to a forward problem component with the target macro-component, the path turns into the FORWARD-PB state.

R6: If a path is in the I state and the results of the fault simulator lead to a backward and forward problem compared with the target macro-component, the path turns into the BACKWARD-FORWARD-PB state.

R7: At the end of the fault simulation, if all path macro-components are tested, the path turns into the T state.

R8: If the path is in the BACKWARD-PB state and there are marked macro-components before the target macro-component or the macro-component is not common with other paths, then path turns into the BACKWARD-C state (i.e. the ATPG will be again applied to the first marked macro-component before the target one.

R9: When the path is in the BACKWARD-C state:

— If there are again backward problems, the path turns into the BACKWARD-PB state.

— If there is no problem, the path turns into the T state.

Other transition rules are given by the expert. They differ from the above nine rules because they constitute what we call the *"path state maintenance"*.

These rules cannot be applied to the situation where the path turned from the C or BACKWARD-C state to the T, BACKWARD-PROBLEM or BACKWARD-FORWARD-PROBLEM state, and also to the situation where the path turned from the C or BACKWARD-C state to the FORWARD-PROBLEM if the forward macro-components are common with other paths.

Maintenance rules

These rules cannot be applied to all paths, but only to those that have common macro-components with the path just treated.

R10: If the path is in the I state, it turns to the TM state.

R11: If the path state is TM, and the list of the macro-component already treated is equal to the list of the macro-components in the path, then the path turns into the T state.

R12: If the path state is BACKWARD-PROBLEM and the macro-component, before the target the macro-component, belongs to the list of the macro-components already treated (LATM), then the path turns into the T state.

R13: If the path state is FORWARD-PROBLEM and the macro-component, after the target macro-component, belongs to the LATM, then the path turns into the T state.

R14: If the path state is BACKWARD-FORWARD-PROBLEM and the macro-component before the target the macro-component, belongs to the LATM, then the path turns into the FORWARD-PROBLEM state.

R15: If the path state is BACKWARD-FORWARD-PROBLEM and the macro-component after the target macro-component, belongs to the LATM, then the path turns into the BACKWARD-PROBLEM state.

R16: If the path state is BACKWARD-FORWARD-PROBLEM and the list of the macro-components in the path is equal to the LATM, then the path turns into the T state.

During the application of the above maintenance rules on the path state, maintenance rules must also be applied to the criteria.

The modification rules corresponding to the various criteria are:

— *For criteria about the type of path*: It must be again calculated taking into account the macro-components already tested belonging to all the paths which have changed state.

— *For criteria about the distance from the inputs*: It must be again calculated if the first marked macro-component belongs to the list of already tested macro-components.

C. Analysis of the result

When a path is selected, its description is given at the gate level, and the test sequences associated to a target-macro-component are given to the fault simulator. For each macro-component, the simulator gives back, the list of faults tested by the simulation.

Our system calculates, for each macro-component of the path, its coverage rate considering the faults tested during the previous simulation. Then, knowing the coverage rate of the macro-components in the path, the path state is deducted by the application of the transition rules. The coverage rate is not sufficient if 70 percent of the macro-component faults are not yet covered. While paths are still in the I or TM states, the modules of *selection*, of *call to the* ATPG, and of the *fault simulation* are activated. When all paths are in states different from I or TM, the different procedures are no more executed. Various conclusions about our problem can the be stated.

When all paths are in the state T, the problem solution is finished and all board macro-components can be tested by the given test sequences. If some paths are, at the analysis end, in a state different from the T state we can generally conclude that the board conception must be expected.

5. FUTURE WORK

Our aim is to extend the work realized at the board level for the elaboration of a test pattern generation system at the computer level.

To this end we must reduce the problem to a set of sub-problems in order to use the existing tools on a smaller set of components.

A computer is described by an interconnection of digital boards which are specified by interconnections of integrated circuits. The test pattern generation must be performed in a hierarchical way, namely:

— test pattern generation at the board level, and

— test pattern generation at the computer level.

A test pattern generation system at the board level has been presented in this present chapter. The test pattern generation at the computer level requires test pattern integration. This problem can be split into:

1. Elaboration of a computer model,
2. Test generation at the board level,
3. Resolution of the controllability and observability problem.

To solve the two end points (first and third points), our system seems to have to reproduce the human expert approach. So the controllability and observability problem must be solved by taking into account only simple behaviour given by the expert. Our system will perform the test generation process in different stages according to the expert knowledge. The principal features are:

— Description and modelling of a computer.

— Construction of a behavioral model of the computer for a given board (based on behavioral simplifying expert rules).

— Symbolic propagation on the previous model.

— Application of the test pattern generation system, presented above, on a target board.

— Analysis of the results at the global level.

6. CONCLUSION

Presently both the hierarchical multi-view modelling and the expert system for test pattern generation at the board level tasks run on a symbolic machine, inplemented in a suitable FRL-like language.

An extension of this environments predicted to take into account:

1. Management and interactions of multi-view models,
2. Generation of test patterns at a computer level.

The realization of the second issue needs the following:

1. Elaboration of a computer model,
2. Test pattern generation at the board level,
3. Resolution of controllability and observability problem.

The use of our general framework will lead us to the development of a set of tools for behavioral test pattern generation for off-the-shelf VLSIs and VBS.

REFERENCES

1. R. Davis et. al., Diagnosis Based on Description of Structure and Function, *AAAI-82*, 137-142 (1982).

2. M. Genesereth, The Use of Hierarchical Models in the Automated Diagnosis of Computer Systems, *Stanford HPP Memo 81-20* (1981).

3. M. Genesereth, Diagnosis Using Hierarchical Design Models, *Proc. AAAI-82*, Carnegie Mellon Univ., Pittsburgh P.A. 278-283 (1982).

4. N. Giambiasi et. al., Method of Generalized Deductive Fault Simulation, *Proc. 17th Design Automation Conference* (1980).

5. C. Oussalah, J. F. Santucci et. al., Thesee: A Knowledge Based System for Test Pattern Generation, *Proc. 1st Europ. Workshop on Fault Diagnostics Reliability and Related-Knowledge Based Approaches*, (S. Tzafestas, M. Singh, G. Schmidt, Eds.) Vol. 2, 145-165 (Reidel, 1987).

6. N. Giambiasi, B. Mc Gee, et. al., An Adaptive and Evolutibe Tool for Describing General Hierarchical Models Based on Frames and Demons, *Proc. 22nd Design Automation Conference*, Las Vegas (June, 1985).

7. B. Kuipers, Commonsense Reasoning About Causality: Deriving Behavior form Structure, *IJCAI 84* and *Artificial Intelligence Journal*, 24:169-203.

8. R. Lbath, N. Giambiasi, C. Oussalah, A General Framework for Hierarchical/Multi-views Modellization, *Proc. IASTED ASM'87*, Santa Barbara Cal. (May, 1987).

9. M. Minsky, A Framework for representing Knowledge, in: Winston, P. *The Psychology of Computer Vision*, McGraw Hill (1975).

10. J. Nestor and D. E. Thomas, Defining and Implementing a Multilevel Design Representation with Simulation Applications, *Proc. 19th Design. Automation Conference* (1982).

11. W. Vancleemput, A Hierarchical Language for the Structured Description of Digital Systems, *Proc. 14th. Design Automation Conference*, 272-279 (1977).

12. M. Stefik and D. G. Bobrow, Object Oriented Programming: Themes and variations, *AI Magazine* (1984).

7

7

ATTRIBUTE GRAMMAR APPROACH TO KNOWLEDGE-BASED SYSTEM BUILDING*

— APPLICATION TO FAULT DIAGNOSIS

George Papakonstantinou and Spyros Tzafestas

Computer Engineering Division
National Technical University of Athens
Zografou 15773, Athens, Greece

1. INTRODUCTION

Attribute grammars were devised by Knuth as a tool for the formal specification of programming languages {2}. Almost concomitant with their development, has been their use in applications which cannot be thought as formal specifications in the strict sense. Recently, attribute grammars were proposed as a tool for knowledge representation and logic programming{5,32}. A theoretical study of the relations between attribute grammars and logic programming is provided in {7}. The use of arbitrarily complex attribute evaluation schemes is also discussed in {7} from the viewpoint of logic program's flow control. However practical implementations of this approach have not so far appeared in the literature. The use of attribute grammars for knowledge representation was firstly proposed in {5}, where it was shown that the parsing mechanism and the semantic notation of attribute grammars can be combined to represent the control knowledge and the knowledge base of logic programs, repsectively. The practical implementation of this approach in situations where the knowledge can be expressed in the form of logic rules was described in {6,32}. Some further theoretical aspects of this last approach are included in {20}. On the application side two problems have so far been considered by the authors via the attribute grammar approach; the first concers a car-fault diagnosis problem using the possibility-necessity inexact reasoning model, and the second treats the fault diagnostic problem of a full adder circuit using the full theorem proving capabilities of an extended attribute grammar model {22,33}.

In this chapter, we provide a complete presentation of the attribute grammar approach to knowledge-based system building, including all the aspects mentioned above. We start with the necessary definitions of grammars in Section 2, and then we describe the attribute grammars in Section 3. Section 4 contains the knowledge representation model via attribute grammars, and Section 5 discusses our implementation of an attribute grammar interpreter. The extension of the above attribute-grammar-based model to logic programming for the implementation of diagnostic expert systems is provided in Section 6. The probabilistic inference model the certainty factors'model, and the possibility-necessity model are included. Section 9 closes with an outline of the additional features required in order for the inference mechanism to obtain full theorem proving capabilities. Finally, Section 7 provides the two application examples mentioned above, i.e. an attribute-grammar-based expert system for car-fault diagnosis, and a dull adder fault diagnostic system based on the faull theorem prover. General comprehensive discussions on the knowledge-based approach to system modelling, diagnosis and control can be found in {28-31}.

2. DEFINITIONS OF GRAMMARS

D1. A vocabulary V is defined as a set of symbols. The set of all strings composed of symbols from V including the empty string ε, is denoted as V*, and V^+ denores $V*-\{\varepsilon\}$.

D2. A grammar G is defined {1} as a quadruple $G=(N,T,P,Z)$ where:

   N ia s finite set of symbols called "nonterminals",
   T is a finite set of symbols called "terminals",
   Z is an element of N called "start symbol",
   P is a finite set of productions (rules) having the form $\alpha \rightarrow \beta$. Here "α"is the left part of the production and "β" the right part, with $\alpha, \beta \varepsilon V*, V=NUT, N \cap T=\emptyset$ (α is assumed to have at least one nonterminal symbol). Capital letters will be used for "nonterminals" and lower-case Greek letters for strings of symbols in V.

D3. A string of symbols α' is said to "directly generate" another string of symblos β' ($\alpha' \Rightarrow \beta'$) for $\alpha', \beta' \varepsilon V*$, if there exist $\alpha_1, \alpha_2, \alpha, \beta \varepsilon V*$ such that $\alpha'=\alpha_1 \alpha \alpha_2$, $\beta'=\alpha_1 \beta \alpha_2$ and $\alpha \rightarrow \beta$ is in P.

D4. A "derivation" is a sequence of strings $\alpha_o, \alpha_1, \ldots, \alpha_n$ (n $\geqslant$0) such that $\alpha_o \Rightarrow \alpha_1, \alpha_1 \Rightarrow \alpha_2, \ldots, \alpha_{n-1} \Rightarrow \alpha_n$, globally denoted as $\alpha_o \overset{*}{\Rightarrow} \alpha_n$, or $\alpha_o \overset{+}{\Rightarrow} \alpha_n$ if n $\geqslant$1.

D5. A "sentence" is defined as a string "s" consisting of only terminal symbols, such that $Z \overset{*}{\Rightarrow} s$.

D6. The language L(G) is the set of all sentences generated by G, i.e.
$$L(G)=\{s \in T* | Z \overset{+}{\Rightarrow} s\}$$

D7. A context-free grammar (C F G) is a grammar in which each production has the form
$$A \rightarrow \chi, A \varepsilon N \text{ and } \chi \varepsilon V*$$
According to Chomsky's hierarchy a CFG is also called a "type 2 grammar".

D8. A "parse tree" (derivation tree, syntax tree) of a sentence sεL(G) is a tree which is formed according to the following rules:

   — The root node has the symbol Z as its label.
   — The leaves[1] of the tree are the terminal symbols of the sentence s.
   — If $X \rightarrow y_1 y_2 \ldots y_k$ is a production used in the derivation of s such that $y_i \varepsilon N$ or $y_i \varepsilon T$ for $1 \leqslant i \leqslant k$, then there exists a subtree in the parse tree of the form.

---

1. We may include the null string Λ in the leaves, which corrends to considering rules of the form $X \rightarrow \Lambda$.

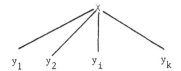

$$y_1 \quad y_2 \quad y_i \quad y_k$$

- Starting from the start symbol Z we construct the parse tree in parallel with the derivation of the sentence s, by attaching to each node, that corresponds to a nonterminal, its descendants according to the previous rule.

D9. A "parser" (parsing algorithm. analyser) for a CFL is an algorithm which decides if a given string is a sentence of the language and constructs parse trees for those strings which are sentences of the language. There are many parsing strategies according to which the parsers can be classified to different categories as follows:

- A top-down parser is a parser that tries to determine the parse tree starting from its root and working down to its leaves.

- A bottom-up parser is a parser that tries to determine the parse tree starting from its leaves and working up to its root.

- A deterministic parser is a parser which at each step can definitely decide if the parsing leads to a succesful solution.

- A non-deterministic parser is a parser which at each step has to guess in order to arrive at a parsing decision. If guessing is proved wrong then backtracking takes place to the most recent guess and another guess is tried. It is obvious that deterministic parsers are more efficient than non-deterministic ones. Not all CFL's can have deterministic parsers.

D10. A CFG G is called "ambiguous" if there exist more than one parse trees for at least one sentence in L(G).

We shall now give an example of a CFG in order to clarify the above definitions. Let us define the unsigned numbers in the binary system, using a CFG. The grammar can be defined as follows:

$$
\begin{aligned}
G &= (N, T, P, Z) \\
\text{where} \quad N &= \{Z, A, K, D\} \\
T &= \{., 0, 1\} \\
P &= \{Z \rightarrow A.K, \\
& \quad Z \rightarrow A \quad, \\
& \quad A \rightarrow AD \quad, \\
& \quad A \rightarrow D \quad, \\
& \quad K \rightarrow DK \quad, \\
& \quad K \rightarrow D \quad, \\
& \quad D \rightarrow 0 \quad, \\
& \quad D \rightarrow 1 \quad \}
\end{aligned}
$$

In Figure 1 the parsing of the string 10.01 is shown.

For simplicity reasons intermediate steps in the construction of the parse tree have been omitted. In representing a CFG, it is customary to use the notation called *Backus-Naur Form* (BNF). The above grammar for the unsigned number is written in BNF form as follows:

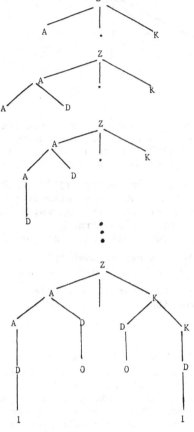

Fig. 1   Parsing of the string 10.01 .

$$<Z>::=<A>.<K>|\ <A>$$

$$<A>::=<A><D>|\ <D>$$

$$<K>::=<D><K>|\ <D>$$

$$D ::= 0|1$$

Each nonterminal is represented by a name (a string of characters) in
angle brackets "< >". The symbol "::=" (or "=") is used for the replace-
ment arrow "→", the symbol "|" is used to separate the alternative re-
placements of the nonterminal to the left of the production.

3. ATTRIBUTE GRAMMARS

An attribute Grammar (AG) {2,3} is based on a context free grammar
$G=(N,T,P,Z)$. Each symbol X in the vocabulary V=NUT of G, has an associated set of attributes A(X). Each attribute represents a specific context
sensitive property of the symbol. The notation X.α is used to indicate
that attribute "α" is an element of A(X). Here A(X) is partitioned into
two disjoint sets, the *synthesized attributes* AS(X) and the *inherited attributes* AI(X). Inherited attributed X.i are those attributes the values
of which are defined in terms of attributes at the parent and/or the
sibling nodes of the corresponding X node in the parse tree. Synthesized
attributes X.s are those attributes whose values are definined in terms
of attributes at descendent nodes of the corresponding X node in the parse
tree.

106

Each of the productions $p \varepsilon P$ $(p:X_o \to X_1 \ldots X_n)$ of the CFG is augmented by a *semantic condition* SC(p) and a set of *semantic rules* SR(p). A semantic condition is a constraint on the values of certain attributes, that are elements of the set $U_{i=0 \ldots n} A(X_i)$, which must be satisfied in each application of the production p. A semantic rule defines an attribute in terms of other attributes of nonterminal X_o on the left hand side of p as well as inherited and synthesized attributes of symbols $X_1, \ldots, X_n$ on the right hand side of p.

More formally an AG is defined as a 5-tuple AG={G,A,D,SR,SC}, where:
G is a CFG {N,T,P,Z}
$A = U_{X \varepsilon TUN} A(X)$
D is the set of domains of all attribute values
$SR = U_{p \varepsilon P} SR(p)$ is a finite set of semantic rules
$SC = U_{p \varepsilon P} SC(p)$ is a finite set of semantic conditions.

The analysis of a string belonging to a language generated by an AG, usually proceeds in two phases:

a) *Syntax analysis*, in which a parser is used to construct s syntax tree from the input string.

b) *Semantic evaluation*, in which the values of attribute occurences at the nodes of the parse tree are evaluated and semantic conditions are tested.

Attribute directed parsing is a technique in which the two phases mentioned above are not distinct, and feedback is allowed from the semantic evaluation phase to the syntax analysis phase. The semantic tree (corresponding to a derivation of a sentence s in AG) is the parse tree of the underlying CFG, "decorated" at each node with the corresponding attribute occurences.

We shall now give an example AG in order to clarify the above definitions. The example is again the definition of the unsigned numbers in the binary system. The meaning (the value of the number) will also be described as well as the constraint that no more than m digits can be used at the fractional part of the number.

Two attributes are introduced namely "val" and "w". The attribute val takes as value the value of the substring recognized so far and the attribute w takes as value the order of the corresponding digit in the franctional part of the string (e.g. -1, -2,...).

Each nonterminal symbol of the grammar has the following attributes:

AS(Z)={val}   AI(Z)={}
AS(Z)={val}   AI(A)={}
AS(K)={val}   AI(K)={w}
AS(D)={val}   AI(D)={}

The AG can be defined as follows:

Productions	Semantic rules and conditions
0: $Z \to A.K$	SR: Z.val:=A.val+K.val; K.w:=-1;
1: $Z \to A$	SR: Z.val:=A.val;
2: $A_1 \to A_2 D$	SR: A_1.val:=A_2.val*2+D.val;
3: $A \to D$	SR: A.val:=D.val;

$$4: \quad K_1 \rightarrow DK_2 \qquad SR: K_1.val:=D.val*2^{K_1.w}+K_2.val;$$
$$K_2.w:=K_1.w-1;$$
$$SC: |K_1.w|<m;$$

$$5: \quad K \rightarrow D \qquad SR: K.val:=D.val* 2^{K.w};$$

$$6: \quad D \rightarrow 0 \qquad SR: D.val:=0;$$

$$7: \quad D \rightarrow 1 \qquad SR: D.val:=1.$$

In the above productions, multiple occurences of the same nonterminal in a production, are identified by indexing. It is observed that the AG has only one semantic condition in the production 4.

If we consider again the input string 10.01 then its semantic tree is shown in Figure 2 (for m=2).

If the value of m was 1, then the input string 10.01 could not be recognized as a sentence of the given grammar, since the semantic condition $|K_1.w|<m$ would not be true.

4. KNOWLEDGE REPRESENTATION WITH ATTRIBUTE GRAMMARS

Knowledge representation is the main activity distinguisting expert system from other application computer systems. Most expert systems organize knowledge on three levels, i.e. *data, knowledge-base,* and *control.* At the data level declarative factual knowledge is represented. For the representation at this level the tools used included first order predicate logic formulas, semantic networks, and frames. At the knowledge-base level inferential knowledge is represented which is necessary for the deduction of new facts not included in the data level {4}. At this last level tha main tools used are *logic programming* (e.g. PROLOG) and *production systems* (e.g. OPS). These two classes of tools represent the two main trends, namely the *declarative* and *procedural* approach to knowledge

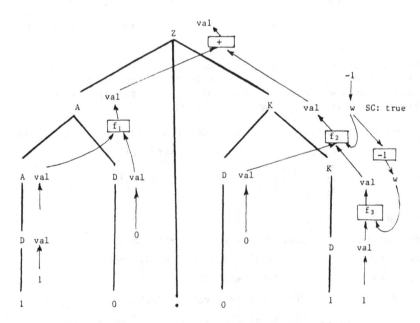

Fig. 2   The semantic tree of the string 10.01 .

representation. The control knowledge defines the inference mechanism which is responsible for the interpretation of the other bodies of knowledge. *Attribute grammars* have been proposed {5-8} as a tool for knowledge representation. Using attribute grammars, we can easily have the two approaches, namely the declarative and the procedural approach, combined in a single tool. Moreover the technology of attribute grammars' processing is fairly mature and many implementations of compilers and interpreters do exist {9,10}.

In the following, we shall use the notation of attribute grammars augmented with a global attribute $FLAG$ which takes the values *true* and *false* {9}. When $FLAG$ takes the value *false* during the semantics evaluation of a BNF rule, the parser considers that matching is not successful. Hence parsing can be directed by the semantics. The above extension is useful for the implementation of the semantic condition feature of AG. The class of attribute grammars to be considered are those which can be evaluated in a single pass from left to right. Bochman {11} has given a condition for an attribute grammar, which assures that the semantics can be evaluated in a single scan from left to right.

The general idea of using an attribute grammar for knowledge representation {5}, is to use only one terminal symbol, the *nil* symbol. Thus the grammar recognizes only *empty strings* of characters. During the recognition of an input string (actually the empty string), using attribute directed parsing, the semantics can be such that at the time they are evaluated they accomplish the inference required.

We shall assume that the knowledge-base can be written in the form of logic rules in a PROLOG like notation:

$$R_0^e(t_{01}^e, t_{02}^e, \ldots, t_{0k_0}^e) \text{ is true IF}$$

$$R_1^e(t_{11}^e, t_{12}^e, \ldots, t_{1k_1}^e) \text{ is true AND}$$

$$\vdots$$

$$R_{m_e}^e(t_{m_e 1}^e, t_{m_e 2}^e, \ldots, t_{m_e k_{m_e}}^e) \text{ is true}$$

$$(1)$$

where t_{ij}^e $(0 \leqslant i \leqslant m_e, 1 \leqslant j \leqslant k_i, 1 \leqslant e \leqslant n)$ is a constant or variable, e is the rule number, and R_i^e is a predicate symbol. If $m_e = 0$ then the rule is a fact, part of the data knowledge. A query which can be a conjunction of R_i^j symbols is considered as a special case of the rule according to (1), where R_o^e is a special symbol "G" that we shall call "start symbol" and which has no parameters. The symbol R_o^e is called the *conclusion* of the rule, while the symbols R_i^e, $1 \leqslant i \leqslant m_e$ are called the *premise items* of the rule and the whole right part of the rule is called the *premise*.

For each logic rule of the form (1), a corresponding syntax rule can be written as well as associated semantic rules for an *"homologous"* (*equivalent*) *attribute grammar*. An homologous attribute grammar is a grammar which when processed by an attribute grammar interpreter will give the same results with the corresponding logic program consisting of rules as in (1).

The corresponding syntax rules (in BNF form) for the logic rule (1) must be written as:

$$\langle R_0^e \rangle ::= \langle R_1^e \rangle \langle R_2^e \rangle \ \ldots \ \langle R_{m_e}^e \rangle \tag{2}$$

For each logic rule (1) the corresponding semantic rules can be written in a mechanical way. To each R_g^e symbol we associate k_g synthesized attributes $R_g^e.s\alpha_j$, $1 \leqslant j \leqslant k_g$ and k_g inherited attributes $R_g^e.i\alpha_j$, $1 \leqslant j \leqslant k_g$. Each pair of these attributes corresponds to each of the k_g arguments of R_g^e, e.g. $R_g^e.s\alpha_j$ means the synthesized attribute of R_g^e corresponding to its jth argument.

Let us illustrate how one can write the semantic rules with the following simple example:

fact : likes (John, Mary)
rule : greets (John, X) if likes (John, X)
query: G if greets (John, ?)   (Whom greets John)

where G is always considered *true*.

The homologous AG can be written as:

syntax rules	**semantic rules**
<G>::=<greets>	1: greets.$i\alpha_1$:="John";
	2: output (greets.$s\alpha_2$);
<greets>::=<likes>	3: *if* greets.$i\alpha_1 \neq$"John" *and* greets.$i\alpha_1 \neq nil$ *then* FLAG:=*false else* greets.$s\alpha_1$:="John";
	4: likes.$i\alpha_1$:="John";
	5: likes.$i\alpha_2$:=greets.$i\alpha_2$;
	6: greets.$s\alpha_2$:=likes.$s\alpha_2$;
<likes>::=*nil*	7: *if* likes.$i\alpha_1 \neq$"John" *and* likes.$i\alpha_1 \neq nil$ *then* FLAG:=*false else* likes.$s\alpha_i$:="John";
	8: *if* linkes.$i\alpha_2 \neq$"Mary" *and* likes.$i\alpha_2 \neq nil$ *then* FLAG:=*false else* likes.$s\alpha_2$:="Mary".

The first semantic rule states that the inherited attribute corresponding to the first argument of the query predicate must take the value "John" while the second semantic rule states that the value of the synthesized attributed of the second argument of the same predicate holds the result. Since the first argument of the predicate greets, which is the conclusion of the second rule, is a constant, it must be unified with the corresponding inherited attribute. This is accomplished with the third semantic rule (actually it is a combination of a semantic rule and a condition). The fourth semantic rule states that the inherited attribute corresponding to the first argument of the predicate "likes" of the premise of the rule must become equal to "John" in order to properly unify with the conclusion of another rule or with a fact. The fifth and sixth semantic rules state that the second argument of the conclusion and the second argument of the predicate likes take the same value. This is accomplished by passing through the corresponding inherited attributes the value of X from the parent to the son in case X already has a value, or passing through the corresponding synthesized attributes the value of X from the son to the parent in case X takes a value later in the inference procedure. The seventh and eighth semantic rules are justified in the same way as the third semantic rule. The semantic tree for the above example is shown in Figure 3.

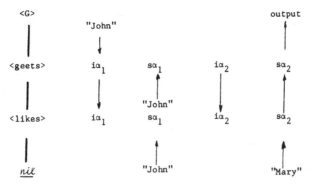

Fig. 3 Semantic tree of the example.

It is obvious that the writing of the semantic rules is a tedious task and so the question of what is the usefulness of this approach may arise. There are three answers to this question:

i) Additional semantics can be easily added using all the facilities offerred in the host language for writing the semantics. With these additional semantics more complex situations can also be handled as will be discussed later.

ii) The technology of attribute grammars is fairly mature and a lot can be gained using the approach proposed.

iii) Most importantly the writing of the semantics can be done mechanically using the information of Table 1 which summarizes all possible cases for each term R_g^e ($1 \leqslant e \leqslant n$, $0 \leqslant g \leqslant m_e$) of a logic rule and for each argument t_{gj}^e ($1 \leqslant j \leqslant k_g$) of R_g^e. Hence this task can be accomplished by a preprocessor {6,32}.

5. IMPLEMENTATION OF AN ATTRIBUTE GRAMMAR INTERPRETER

The interpreter to be described consists of a *parser* representing the *control knowledge*, and an *attribute evaluator* representing the *factual knowledge* and the *inferential knowledge*. The parser is a nondeterministic top-down one which, besides parsing, manipulates a push down stack representing the parse tree and the attribute instances, and calls the attribute evaluator passing to it the appropriate-parameters. The attribute grammar interpreter, based on the above parser and attribute evaluator, performs the parsing and the semantics evaluation simultaneously, so that the parsing can be directed by the semantics. It can accept attribute grammars which can be evaluated in a single pass from left to right.

The attribute grammar interpreter has been implemented in FORTRAN (as well as in BASIC). It consists of about 300 instructions. The code of the attribute evaluator is not included in this number, because its size depends upon the specific application. The attribute evaluator can be manually coded in a quite mechanical way or can be generated automatically by an appropriate preprocessor. Such a preprocessor has been developed when the interfacing language is a subset of PROLOG and consists of about 400 BASIC instructions. The code that it generates is BASIC code.

The corresponding syntax rule (when the preprocessor is not utilized) for the logic rule (1) must be written as in (2):

Table 1 Semantics for the argument t_{gj}^e of R_g^e

	CONDITION	SEMANTIC RULES	POSITION
t_{gj}^e is a constant c and g=0		*if* $ia_j(R_g^e) \neq c$ *and* $ia_j(R_g^e) \neq nil$ *then* flag:= *false else* $sa_j(R_g^e):=c$	immediately after R_1^e
t_{gj}^e is a constant c *and* g≠0		$ia_j(R_g^e):=c$	after R_g^e
t_{gj}^e is a variable *and* g=0		*if* t_{gj}^e is the only occurrence of the variable in the rule, *then* $sa_j(R_g^e):=$ $ia_j(R_g^e)$ *else if* t_{qi}^e is the last textual occurence of the same variable in the production e *then* $sa_j(R_g^e):=$ $sa_i(R_q^e)$	after E^e
t_{gj}^e is a variable *and* it is the first textual occurrence of this variable *and* g≠0		—	—
t_{gj}^e is a variable *and* t_{qi}^e is the nearest textual occurrence of the same variable to the left of t_{gj}^e *and* q=0 *and* g≠0		$ia_j(R_g^e):=$ $ia_i(R_q^e)$	after R_g^e
t_{gj}^e is a variable *and* t_{qi}^e is the nearest textual occurrence of the same variable to the left of t_{gj}^e *and* q≠g *and* g≠0		$ia_j(R_g^e):=$ $sa_i(R_g^e)$	after R_g^e
There are many occurrences of the same variable in a relation R_g^e *and* g≠0		*if* all variables have not the same value or those which are different have not the *nil* value, *then* flag:= *false else* make those having *nil* value, have the value of the others	immediately after the label next to the R^e label (R_{g+1}^e or E^e)

112

$$\langle R_o^e \rangle = \langle R_1^e \rangle \langle R_2^e \rangle \ldots \langle R_{me}^e \rangle \mid \dashv \qquad (2')$$

where the combination of the last two characters indicates the end of the syntax rule. The use of two characters facilitates the parser. If we have p syntax rules with the same left part R_o and with the same number of t_{ij} parameters in the corresponding R_o symbol then these syntax rules must be written in a combined syntax rule of the form:

$$\langle R_o^r \rangle = \langle R_{11}^r \rangle \langle R_{12}^r \rangle \ldots \langle R_{1q_1}^r \rangle \mid \ldots \mid \langle R_{p1}^r \rangle \ldots \langle R_{pq_p}^r \rangle \mid \dashv \qquad (3)$$

Our parser is based on the parser proposed by Floyd {12}. However, it differs from Floyd's parser in the following aspects:

— It does not utilize terminal symbols.
— It employes an extended stack in which it saves the attribute values as well.
— It includes calls to the attribute evaluator (subprogram EVAL), passing appropriate parameters.
— It uses a meta-variable named "FLAG" which is utilized by the attribute evaluator and its value affects the parsing.
— It finds all possible parse trees.

The parser can informally be described as follows. Let us consider the syntax rule of (3). It first attempts to recognize the R_{11}^r nonterminal symbol trying this subgoal. When the control returns to R_o^r with success reported, then subgoal R_{12}^r is tried, then R_{13}^r and so on. If the symbol "$\mid$" is reached in this way R_o^r reports success to his parent. When the control returns to R_o^r after trying the subgoal R_{1j} and failure is reported then the predecessor of R_{1j}^r i.e. R_{1j-1}^r is tried again if $j>1$. If $j=1$ then the next alternative is tried. If no other alternatives can be tried then R_o^r reports failure to his parent.

The parsing that takes place is degenerate since there are no terminal symbols. Hence, success must always be reported, provided the non-terminal symbols are all defined. However, tha parser has been extended to incorporate calls to an attribute evaluator procedure called EVAL, just before trying each subgoal R_{ij}^r, in order to evaluate the inherited attributes of R_{ij}^r. It also calls EVAL after succeeding in the last subgoal $R_{im_i}^r$ of the alternative, in order to evaluate the synthesized attributes of R_o^r (parent node). The evaluation of the attributes of R_{ij}^r corresponds to the unification procedure as will be explained in the next section. Hence, it is possible that in the semantics (evaluation of attributes) some conditions will not hold. This will be considered as failure reported by the subgoal R_{ij}^r. The metavariable FLAG is used for this purpose, which is tested in the parser after the control returns to R_o^r.

The parser, will now be formally described using the notation of {12}.

Let: S   be a stack of six-tuples $(goal_\lambda, i_\lambda, sup_\lambda, sub_\lambda, pred_\lambda, A_\lambda)$.

λ be the element of the stack which is currently active.

$goal_\lambda$ be the nonterminal symbol which is currently tried in the parsing.

i_λ be the place in the definition of the nonterminal goal at which parsing has reached so far.

sup_λ be the location in the stack S of the $goal_\lambda$'s superior (parent).

sub_λ be the location in the stack S of the $goal_\lambda$'s most recent subordinate (descendent).

$pred_\lambda$ be the location in the stack of the $goal_\lambda$'s most recent sibling.

A_λ be the attribute occurences set to be defined in the next section.

v be the first free element of the stack S.

Each word in the stack represents a node in the parse tree, where goal represents the label of the node, i is the index in the grammar showing where the scanning of the rule applied in the node has reached so far, sup designates the parent node, sub designates the rightmost son of the node, and pred designates the sibling immediately to the left of the node. As an example, the grammar

    `<G>=<A><B>| ⊣`
    `<A>=<C>| ⊣`
    `<G>=| ⊣`
    `<B>=| ⊣`

will at the end create the stack

	Goal	i	sup	sub	pred
1	G	5	0	4	0
2	A	10	1	3	0
3	C	14	2	0	0
4	B	18	1	0	2

We note that each nonterminal is counted as one symbol, and the grammar is considered as a continuous string of symbols ignoring the blanks.

The flowchart of the parser is shown in Figure 4. As another example, consider the grammar

    `<G>=<A><B>| ⊣`

    `<A>=<B><C>|<C><D>|⊣`

    `<B>=<D>|<E>|⊣`

    `<C>=<E>|⊣`

    `<D>=|⊣`

    `<E>=|⊣`

The various parse trees which will be generated are shown in Figure 5.

Since the attribute grammar can be evaluated in a one pass evaluator, the attributes will be evaluated in the following sequence:

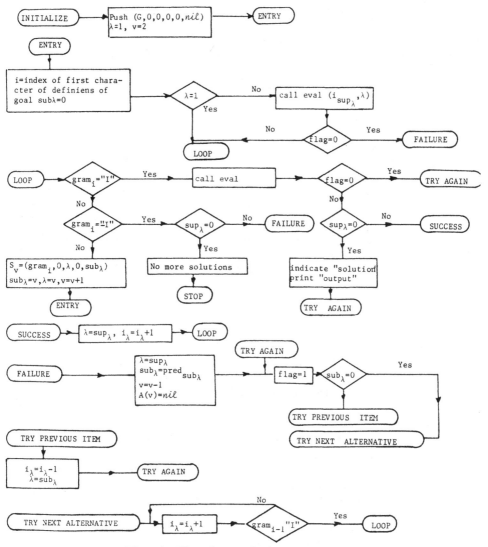

Fig. 4 Flowchart of the parser.

$$ia(R_o),\ i_a(R_1),\ sa(R_1),\ ia(R_2),\ sa(R_2),\ldots,ia(R_m),\ sa(R_m),\ sa(R_o)$$

The attributes sa and ia for R_g^e are used to form the attribute occurences set A_λ defined as:

$A_\lambda=\{sa(\lambda,j),\ ia(\lambda,j)\ 1\leqslant j\leqslant k_g$ and λ as defined in the previously section} (4)

Here, 2 * max memory locations have been reserved as part of each element in the stack, where max is the maximum number of parameters of any predicate symbol R_g^e appearing in the logic program. An attribute $xa_j(R_g^e)$ where x stands for either s or i, occupies a known position in the stack S, provided the position λ of the instance of R_g^e is known,or the position of one of its relatives (parent, son, sibling) is known. Such information is passed to EVAL by the parser in the second argument of the corresponding call. A function $f(p)$ can then be evaluated using additional information in the stack S such that:

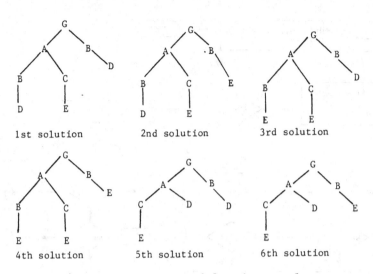

Fig. 5 The parse trees generated for the example grammar.

f(0) gives the position in the stack of the parent node
f(1) gives the position in the stack os the first son
f(2) gives the position in the stack of the second son etc.

Using this function, the position in the stack of an attribute occurrence of $xa_j(R_g^e)$ (x stands for either s or i) will be $xa(f(g), j)$. This last symbolism will be used in the EVAL when writing semantic rules that are constructed from the logic program according to Table 1. In Table 1 the place of the semantic rules in EVAL is also given. The procedure EVAL must have the following structure:

procedure EVAL(i,j);
begin integer i,j;
    case i of
        ⋮

Repeat   position in grammar of the first son R_1^e: semantic rules
for each   position in grammar of the second son R_2^e: semantic rules
                    ⋮
corresponding
                position in grammar of the last son R_{me}^e: semantic rules
logic rule   position in grammar of the end of the alternative E^e:
e   semantic rules
                    ⋮

    end case
        ⋮
end EVAL

The parser calls EVAL and passes in the first argument of the call the position in the grammar of the nonterminal which is going to be tried next. Hence in the EVAL the appropriate semantics for evaluating the inherited attributes of this nonterminal will ne executed. The parser also calls EVAL at the end of an alternative passing of this position in the grammar, in the first argument. Hence in EVAL the appropriate semantics for evaluating the synthesized attributes of the parent node will be executed.

We can see that when the logic rules are given, EVAL can be mechanically constructed with the help of Table 1. This is the only work a designer has to do in order to build a specific expert system with this tool. However, as we have mentioned before this task can be done automatically using a preprocessor. The preprocessor accepts as input the logic rules and produces EVAL. Additional semantic rules can also be inserted by the user.

Suppose that we have the following knowledge base:

successor (X,Y) if parent (Y,Z) and successor (X,Z)   (5)

successor (X,Y) if parent (Y,Z)   (6)

parent (j,b)   (7)

parent (j,ℓ)   (8)

parent (b,a)   (9)

parent (b,p)   (10)

and we want to ask the question

p is successor of whom? i.e. successor (p,?)

The syntax rules of the equivalent grammar are:

$\langle G \rangle = \langle successor \rangle \, | \dashv$ (11)

$\langle successor \rangle = \langle parent \rangle \langle successor \rangle \, | \langle parent \rangle \, | \dashv$ (12)

$\langle parent \rangle = |\,|\,|\,| \dashv$ (13)

Rule (11) is the generic rule corresponding to the query. Rule (12) has two alternatives corresponding to (5) and (6) respectively, and rule (13) has 4 alternatives corresponding to (7), (8), (9) and (10). Each alternative of rule (13) is *nil* since (7), (8), (9) and (10) are facts.

Using Table 1 and the rules (5)-(10) we can now write the procedure EVAL as follows:

```
procedure EVAL (p,q);
begin integer p,q;
    case p of
 3: ia(f(1),1):="p";
 4: print(sa(f(1),2));
 8: ia(f(1),1):=ia(f(0),2);
 9: ia(f(2),1):=ia(f(0),1);
10: sa(f(0),1):=sa(f(2),1);
    sa(f(0),2):=sa(f(1),1);
11: ia(f(1),1):=ia(f(0),2);
    ia(f(1),2):=ia(f(0),1);
12: sa(f(0),1):=sa(f(1),2);
    sa(f(0),2):=sa(f(1),1);
16: if ia(f(0),1)≠"j" and ia(f(0),1)≠nil then FLAG:=false else sa(f(0),1):="j";
    if ia(f(0),2)≠"b" and ia(f(0),2)≠nil then FLAG:=false else sa(f(0),2):="b";
17: if ia(f(0),1)≠"j" and ia(f(0),1)≠nil then FLAG:=false else sa(f(0),1):="j";
    if ia(f(0),2)≠"ℓ" and ia(f(0),2)≠nil then FLAG:=false else sa(f(0),2):="ℓ";
18: if ia(f(0),1)≠"b" and ia(f(0),1)≠nil then FLAG:=false else sa(f(0),1):="b";
    if ia(f(0),2)≠"a" and ia(f(0),2)≠nil then FLAG:=false else sa(f(0),2):="a";
19: if ia(f(0),1)≠"b" and ia(f(0),1)≠nil then FLAG:=false else sa(f(0),1):="b";
    if ia(f(0),2)≠"p" and ia(f(0),2)≠nil then FLAG:=false else sa(f(0),2):="p";
end case;
end EVAL
```

It has been assumed that the syntax rules are forming a continuous string of symbols and each nonterminal is sonsidered as a single symbol. The answer will be given with the sementic rule print(sa(f(2),2)) after the label 4 of EVAL, which has been added in order to print the evaluated value of sa_2(<successor>) of rule (11) which corresponds to the query.

The stacks in their final form for the two solutions found are given in Table 2. The corresponding parse trees are given in Figure 6.

6. EXTENSIONS

In this section the previous attribute grammar-based approach will be extended to logic programming, and it will be explained how this approach can be used for the implementation of expert systems suitable for diagnostic purposes. We shall mainly concentrate on the means of implementing features particularly useful in diagnostic tools such as inference generated requests for the establishement of facts, explanation of inference process flow, reasoning under uncertain beliefs, management of the imprecision of the established facts and finally full theorem proving capabilities.

6.1. Inference Generated Requests

When the establishment of a fact is sought by the inference engine, according to the assertions contained in the knowledge base and this trial fails, it is useful that the system asks the user to verify the sought fact. This can be successfully accomplished using the formalism to attribute grammars. Each time it is verified that an atome cannot be automatically deduced, a semantic function can be activated, which asks the user to verify this atom. If he does, the atom is in the sequel considered as established.

6.2. Why Explanations

When the user is faced against an inference generator request for data, as described just before, it is reasonable for him to ask why this data is being requested. To be able to respond to such a "why" request, the system should be able to print the rule whose the atom requested to be verified, is a premise. To do this in the formalism of attribute grammars, each nonterminal symbol (which in fact corresponds to an atom) can be assigned an attribute, whose value is a pointer to a record of a text file. This record contains the text that should be given as answer, when a why explanation is requested during the inference generated request for the establishment of the corresponding atom. Each time the system is asked for such an explanation, a global semantic function can be invoked, which prints the text corresponding to the current value of the text file pointer attribute. It should be noted that this attribute is imposed or built-in, i.e. its value is fixed and it doesn't come from the context of the corresponding nonterminal symbol within the grammar.

6.3. How Explanations

The explanation ability of the reasoning process is one of the most important features of a diagnostic expert system. Each time a diagnosis is produced, the user should be convinced that this diagnosis is reasonable. Moreover, when the knowledge base of such an expert diagnosis system is under debugging, the system's programmer should be able to trace the inference making process and locate the bugs. So, the system must be able to provide "how" explanations, i.e. to give, at the end of the reasoning process, a means of account on how it reached its conclusions.

Table 2 Final form of the stack

Goal	i	sup	sub	pred	attribute occurences			
					$i\alpha_1$	$s\alpha_1$	$i\alpha_2$	$s\alpha_2$
G	4	0	2	0				
SUC	10	1	4	0	P	P		J
PAR	16	2	0	0		J		B
SUC	12	2	5	3	P	P	B	B
PAR	19	4	0	0	B	B	P	P

Goal	i	sup	sub	pred	attribute occurences			
					$i\alpha_1$	$s\alpha_1$	$i\alpha_2$	$s\alpha_2$
G	4	0	2	0				
SUC	12	1	3	0	P	P		B
PAR	19	2	0	0		B	P	P

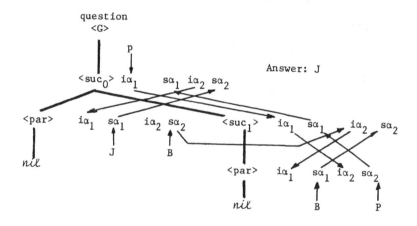

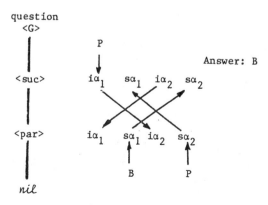

Fig. 6 Parse trees of the example.

To increase the robustness of this sort of explanation, we consider the "how" explanation request to be of the form:

$$\text{how } R_i(t_{i1}, t_{i2}, \ldots, t_{ik})$$

where the variables of the atom R_i may have been substituted by constants.

When such a request is made, a routine is activated which locates the position of the atom R_i within the parse tree, and finds the atom in which it is analyzed. The user can then repeat the above procedure for subsequent atoms. Hence, the diagnostics engine is capable of justifying its conclusions.

6.4. Probabilistic Inference

Let

$P(B_i)$ be the a priori probability of B_i, and

$P(A/B_i)$ be the posterior probability of A, given that B_i is now an item of evidence.

Then, by Bayes's Theorem,

$$P(A) = \sum_i P(A/B_i)P(B_i) \tag{14}$$

The summation applies over all the items of evidence B_i, which must be exhaustive and mutually exclusive. In the following, the stochastic attribute context-free grammars {13} will be tailored to meet the requirements of our specific application.

A stochastic attribute grammar is an AG where to each rule a conditional probability is associated, which is the conditional probability of the conclusion, given its premise.

Consider a rule of the form:

$$\langle R_o^e \rangle ::= \langle R_1^e \rangle \langle R_2^e \rangle \ldots \langle R_{m_e}^e \rangle$$

To each term R_i^e we assign a synthesized attribute p, namely $R_i^e.p$, symbolized by p_i^e for simplicity, which holds for its probability. Furthermore, we assign an additional built-in attribute to R_o^e, namely $R_o^e.p$ symbolized by p_e for simplicity, which stands for the probability associated to the rule, following the definition of the stochastic attribute grammars. Then the corresponding semantic rule for the evaluation of the probability $R_o^e.p$ will be:

$$p_o^e := p_e * p_1^e * p_2^e * \ldots * p_{m_e}^e$$

When more than one paths lead to the same conclusion, Bayes' theorem (14) is used to evaluate the probability of the conclusion.

The semantic rules for the evaluation of the probabilities are very simple and of the same form for all the rules. They are just an implementation of simple probability calculus using synthesized attributes. One more feature can be incorporated, using the metavariable FLAG, mentioned previously. Optionally, a lower threshold may be defined for the probability of an item of evidence. If the probability of occurence of this item is lower than the specified threshold, FLAG becomes false and a backtracking procedure is initiated, thus resulting to an elimination of the low likelihood conclusion.

It is obvious that the parser described in the previous section can not find all paths leading to the same conclusion simultaneously. A well known parser proposed by Early {14} is more suitable for this purpose.

It is reminded that this approach violates the axioms of probability, when the premises are not mutually exclusive or the premise items are not independent. It is evident that it is the user's responsibility to supply a consistent knowledge base, with respect to probability assignment. Otherwise, the results will give a measure of certaintly, but not probabilities, in the strict sense. Because of this disadvantage, the certainty factors'algebra used in MUCIN can be implemented using attribute grammars.

6.5. Certainty Factors

As in MYCIN {15} each rule is assigned an attenuation coefficient (AC) which is a real number between 0 and 1, representing the credibility to the rule. Moreover, each fact is assigned a certainty factor (CF) which is a real number between −1 and 1 and represents a measure of the belief (positive value) or disbelief (negative value) to the fact. Furthermore, when a rule produces a certainty factor CF1 (assumed $\geqslant 0$) for a fact and another rule produces a certainty factor CF2 (assumed $\geqslant 0$ too) for the same fact, then the "accumulated certainty factor", taking into the information that arised from the applications of the two rules, is defined by

$$CF=CF1+CF2-CF1*CF2 \tag{15}$$

This results in a reinforcement of certainty, as the certainty factor is moved asymptotically towards unity, as supporting evidence piles up.

Furthermore, an analogous formula is used when the two combined certainty factors are negative:

$$CF=CF1+CF2+CF1*CF2 \tag{16}$$

When the two combined certainty factors are of opposite sign, the combination formula is:

$$CF=(CF1+CF2)/\{1-\min(|CF1|, |CF2|)\} \quad \text{if } |CF1|*|CF2|\neq 1$$
$$CF=1 \qquad\qquad\qquad\qquad\qquad \text{if } |CF1|*|CF2|=1 \tag{17}$$

Moreover, premises "in conjunction" result in the assignment of the minimum among their certainty factors, while premises "in disjunction" result in the assignment of the maximum among their certainty factors, to the certainty factor of the combined premise.

An additional feature of MYCIN's plausibility evaluation scheme is that, when the application of all the rules which entail a proposition P, produce a certainty factor between −0.2 and 0.2, the inference engine sets the certainty factor of P equal to zero. In the implementation described here, instead of using a fixed value as a threshold, this threshold is an externaly supplied variable, which is hereafter denoted by trsh.

The main principle for certainty factor (CF) evaluation is that the certainty factor of the conclusion of a rule is the product of the certainty factors of the combined premise and the attenuation coefficient (AC) of the rule.

Having in mind these principles, one can easily proceed to the semantic rules for the evaluation of the certainty factors. Consider the rule:

$$<R_0^e> ::= <R_1^e><R_2^e>...<R_{m_e}^e>$$

We associate a synthesized attributed c with each predicate of the rule, to represent its certainty factor. Let these attributes be denoted as $c_0^e, c_1^e, ..., c_{m_e}^e$. Moreover a built-in attribute a_e is used to denote the attenuation coefficient of the rule (this now is the value of p_e, associated with the rule, in the formalism of stochastic attribute grammars). Then, the semantic rule for the evaluation of c_0^e, given that $c_1^e, c_2^e, .., c_{m_e}^e$ have taken values by previous inference, is

$$c_0^e := a_e * \min(c_1^e, c_2^e, ..., c_{m_e}^e) \tag{18}$$

under the assumption that no previous evidence about R_0^e existed in the knowledge base. However, taking into account that formula (15) gives CF=CF1 when CF2=0, we can assign initial values of zero to all c_j^e, except for those corresponding to facts. A value of zero to a certainty factor represents total ignorance on the corresponding fact and hence does not influence the inference. Then, the semantic rule (18) must be modified as follows:

```
        begin flag:=true;
ch:         for i:=1 to m_e do
            if c_i^e<=trsh then flag:=false;
        if flag= true then
            begin c_aux:=a_f^e*min(c_1^e,c_2^e,...,c_{m_e}^e);
            if c_0^e*c_aux<0 then
                begin
                if abs(c_0^e*c_aux)=1 then c_0^e:=1
                    else c_0^e:=(c_0^e+c_aux)/(1-min(abs(c_0^e), abs(c_aux)))
                end
            else if c_0^e<0 then c_0^e:=c_0^e+c_aux+c_0^e*c_aux
                else c_0^e:=c_0^e+c_aux-c_0^e*c_aux
        end
```

The reader should note that the statement labeled (ch:) in the semantics given, disables the use of a fact with certainty factor absolutely less than the specified threshold trsh. The label (ch:) has been added for reference purposes only.

A feature of MYCIN that is important to be implemented using the formalism of attribute grammars is that of autoreferencing rules. An autoreferencing rule is a rule that concludes to one of his premise with a certainty factor substantially greater. For example:

if X reads Reader's Digest
and it seems possible (CF>0.3) that he will vote Republican
then there is a strongly suggestive evidence (CF=0.75) that X will vote Republican.

The syntactic rule corresponding to this logic rule, if <RG> stands for

"X reads Reader's Digest" and <REP> stands for "X will vote Republican", is

$$<REP_0>::=<RG_1><REP_2>$$

It is evident that in such a case, a recursive triggering of this rule, starting from a low certainty factor, say 0.22, will repidly increase the certainty, although no additional evidence has been cumulated. In order to eliminate the risk of such an irrelevant increase of the certainty, the metavariable FLAG which was used to guide the parsing according to the unification of variables, can also be used here. For this purpose, the following semantic rule can be used, its evaluation being accomplished along with that of the inherited attributes of <REP>:

$$\text{if } c_0 < 0.3 \text{ or } c_0 > 0.75 \text{ then FLAG} := \underline{false}$$

This condition will disable the triggering of the rule until other rules increase the certainty factor c_0 to a value greater than or equal to 0.3.

Also, the rule cannot be triggered if the CF is greater than 0.75, because in such a case a recursive triggering would increase iteratively the certainty factor. Then, if <RG> has been established, the rule will be triggered and the certainty factor will be updated using a very simple semantic rule, evaluated along with the synthesized attributes of <REP>:

$$c_0 := 0.75$$

These hints on the treatment of auto-referencing rules have outlined a very interesting property of the proposed system for plausible inference using attribute grammars: the programmer of an expert system can add his own semantic rules to tailor the chaining and affect the inference according to the requirements of his specific application.

In order to avoid the task of writing the semantic rules for each logic rule, appropriate procedures can be incorporated in the preprocessor which writes the semantic rules used for the unification of variables. These procedures take as input the values of the built-in attributes and the probabilities of the facts in the case of Bayesian inference. When the MYCIN's plausible inference scheme is used, the attenuation coefficients of the rules and the certainty factors of the initial facts are given. The output of this preprocessor is a complete procedure for attribute evaluation. In this procedure, the semantic rules that guide the unification of the variables and the plausible inference are incorporated. Hence, the system's programmer has just to supply the logic rules and the facts in a PROLOG-like notation and those among the constants affecting the plausible inference which form part of his evidence. All the rest is absolutely automated.

6.6. The Possibility-Necessity Model

Alternatively, instead of syppling a precise measure of certainty for each item of evidence, an *upper* and a *lower bound* of the corresponding probability namely a *possibility* and a *necessity measure* respectively, can be supplied {16,17,18}. In this case, the scheme for the evaluation of the *possibility* and *necessity* of the conclusion of a rule is as follows

$$N(p \to q) \geqslant a \qquad \Pi(p \to q) \geqslant A \qquad \{\max(1-a,A)=1\}$$

$$N(p) \geqslant b \qquad \Pi(p) \geqslant B \qquad \{\max(1-b,B)=1\}$$

$$N(q) \geqslant \min(a,b) \qquad \Pi(q) \geqslant \max(A.v(A+b), B.v(a+B))$$

Here N and Π denote the *necessity* and the *possibility*, respectively, while

$$v(x+y) = \begin{cases} 1 & \text{if } x+y>1 \\ 0 & \text{if } x+y<1 \end{cases}$$

In practice, the above inequalities may be replaced by equalities. The possibility and necessity measures should satisfy the equality

$$\max\{\Pi(q), 1-N(q)\}=1$$

As this is not always assured, the following normalization should be aplied (the symbol $*$ denotes normalized measures).

$$\Pi^*(q) = \frac{\Pi(q)}{\max\{1-N(q), \Pi(q)\}}$$

$$N^*(q)=1 - \frac{1-N(q)}{\max\{(1-N(q), \Pi(q)\}} \tag{19}$$

The above scheme can also be implemented using synthesized attributes for the *possibility* and the *necessity*.

Consider again the rule

$$<R_0^e>::=<R_1^e><R_2^e>\ldots<R_{m_e}^e>$$

We associate two synthesized attributes with each R_i^e, one for the possibility of R_i^e namely p_i^e and one for the necessity namely n_i^e. Moreover two built-in attributes p_e and n_e are used to denote the possiblity and the necessity of the rule i.e. of the conclusion given its premise. The possibility and the necessity of the conjunction of the R_i^e, $1 \leq i \leq m_e$, symbols are taken respectively as:

$$\Pi_e := \min(p_1^e, p_2^e, \ldots, p_{m_e}^e) \quad \text{and}$$
$$N_e := \min(n_1^e, n_2^e, \ldots, n_{m_e}^e)$$

Hence the possibility and necessity of R_0^e can be estimated {19}, {20} as lower bounds

$$p_0^e = \max(p_e \cdot v(p_e+N_e), \ \Pi_e \cdot v(n_e+\Pi_e)) \quad \text{and}$$
$$n_0^e = \min(n_e, N_e) \text{ respectively.} \tag{20}$$

When more than one premise combinations lead to the same conclusion, the certainty measures coming from different paths, can be combined using Dempster's rule {16}.

6.7. Imprecise Reasoning

When the attribution of a parameter is fuzzy, i.e. the parameter is known to have a value in a certain domain, while the exact calue is not known, one can describe this attribution in terms of a possibility distribution on a domain of possible values. The value of this distribution at each particular point of the domain, represents an upper bound of the probability that the parameter under consideration has the value corresponding to this point.

When the fuzzy event A is the premise of a logic rule, while another fuzzy event A' is the fact invoked as a means of satisfaction of the premise A, and the corresponding possibility distributions are μ_A and $\mu_{A'}$, respectively, over the same domain S, then the degree of satisfaction of the premise is expressed in terms of possibility and necessity measures, according to the formulae {21}:

$$\Pi(A;A')=\sup_{s \varepsilon S} \min\{\mu_A(s),\ \mu_{A'}(s)\}$$

$$N(A;A')=\inf_{s \varepsilon S} \max\{\mu_A(s),\ 1-\mu_{A'}(s)\}$$

These certainty measures are, in the sequel of the reasoning process, considered as the possibility and the necessity of the premise of the rule under consideration. If $\Pi(A')$ and $N(A')$ have already been calculated for the event A' (as values of synthesized attributes), then the following formulae could be applied:

$$\Pi(A)=\min\{\Pi(A;A'),\ \Pi(A')\}$$

$$N(A)=\min\{N(A'A'),\ N(A')\}$$

For further details, the reader may refer to {17} {20}.

6.8 Full Theorem Proving Capability

The kind of rules and the inference mechanism described in sections 4 and 5 can not form a complete theorem proving system. Additional features have to be added for this purpose. These features can be summarized {22}, {23}, {24}, {33}, as follows:

1. The arguments of the predicate symbols R_i^e should be more general, i.e. "terms" as defined in the first order predicate calculus.

2. The occurs test should taken into account.

3. The reduction operation {22} should be incorporated for the inference, in addition to the normal extension operation used in PROLOG and in the system described perviously.

4. The search strategy should be changed to a bounded or staged depth-first search strategy.

It is not very difficult to incorporate the above extensions to the system described so far { 25,32,33}. It is beyond the scope of this chapter to describe all the details. The only feature that will be further discussed is the third one which is more important.

PROLOG uses Horn clauses which are first order predicate calculus clauses that contain at most one positive literal, i.e. clauses of the form:

$$R_0 \lor \sim R_1 \lor \sim R_2, \ldots, \lor \sim R_n \qquad (21)$$

where the symbol "$\sim$" means "NOT" and the symbol "$\lor$" means "OR". The above clauses are written in PROLOG notation as:

$$R_0 :- R_1 \land R_2 \land \ldots \land R_n \quad \text{or} \quad R_0 = R_1, R_2, \ldots, R_n \qquad (22)$$

where the symbol "$\land$" means "AND". In order to be able to have a complete inference system in PROLOG-like systems, we should take into consideration, all the equivalent forms of the form (22) obtained from (21) i.e.:

$$\sim R_1 :- \sim R_0 \land R_2 \land R_3 \land \ldots \land R_n$$

$$\sim R_n :- \sim R_0 \land R_1 \land R_2 \land \ldots \land R_{n-1} \qquad (23)$$

The rules (23) are called general contrapositives of the rule (22).

The reduction operation can now be stated as follows: if the current goal matches the complement of one of its ancestor goals, then apply the matching substitution and treat the current goal as if it were solved. Hence the goal P can be proved from the PROLOG-like rules P:-Q∧R, Q:-∿P and the fact R.

Concluding this section we can state that using attribute grammars we can combine the declarative approach (e.g. PROLOG) and the procedural approach (e.g. OPS) into a single tool. Moreover using additional semantics we can handle characteristics like the ones described in this section or even incorporate ad hoc emperical knowledge into our system.

7. EXAMPLES

In this section we shall describe two examples. The first concerns a toy-scale expert system for car-fault diagnostics using the possibility-necessity model {19}. The second concerns a fault diagnostic procedure for a full adder circuit, using the extensions described in the previous section for full theorem proving capabilities {22}, {26}, {33}.

Example 1: An expert system for car-fault diagnostics

Since the logic rules are very simple, we shall directly give the synthax rules of the homologous attribute grammar. Next to each syntax rule, we give the values of the *possibility* and the *necessity* of the conclusion of the rule, given that its premises hold true:

<goal>::=<car out of order>   1.0   1.0

<car out of order>::=
        <flat battery>/   1.0   1.0
        <fault in ignition commutators>/   1.0   1.0
        <broken camshaft driving chain>/   1.0   1.0
        <fault in carburetory>/   1.0   1.0
        <no gasoline>   1.0   1.0

<flat battery>::=   0.97 0.0
        <no lights working><no starter
        motor turning>

<no gasoline>::=   0.95 0.0
        <gasoline gauge low>
        <no flat battery>
<no flat battery>$_0$::=<flat battery>$_1$

          if $\Pi_1 < 0.15$ then begin

                  $\Pi_0 := 1 - \Pi_1$; $N_0 := 0$;

                  if $\Pi_0 := 1$ then $N_0 := 1$

                  end

              else FLAG:=false;

(this semantic rule simply means that the battery is considered as not being flat when the possibility of the battery being flat is less than 0.15)

<fault in ignition commutators>::=   0.78   0.0
        <gasoline gauge high>
        <car not recently serviced>
        <high voltage circuit out of order>
        <no flat battery>

```
<high voltage circuit out of order>::=        0.95  0.0
      <spark not produced>
      <clear spark plugs>


<broken camshalf driving chain>::=        0.40  0.0
      <starter motor turning>
      <gasoline gauge high>
      <high voltage circuit in order>
      <car doesn't start>
```

$<$high voltage circuit in order$>_0::=$

   $<$high voltage circuit out of order$>_1$

  $\underline{\text{if}}\ \Pi_1<0.2\ \underline{\text{then}}\ \underline{\text{begin}}$

$$\Pi_0:=1-\Pi_1;\ N_0:=0;$$
$$\underline{\text{if}}\ \Pi_0:=1\ \underline{\text{then}}\ N_0:=1$$

    $\underline{\text{end}}$

     $\underline{\text{else}}\ \text{FLAG}:=\text{false};$

```
<fault in carburetor>::=        0.65  0.0
      <starter motor turning>
      <gasoline gauge high>
      <high voltage circuit in order>
      <car doesn't start>
```

It is noted that $\Pi_i(N_i)$ is the possibility (necessity) of the nonterminal in the rule with index i.

  The semantic rules for the evaluation of the possibility and necessity measures of the conclusion have not been given, because of their simplicity. Suppose now that the user has provided the following facts, as part of his evidence, along with the corresponding certainty measures.

<gasoline gauge high>	1.0	1.0
<car not recently serviced>	1.0	0.5
<spark not produced>	1.0	1.0
<clear spark plugs>	1.0	0.9
<no lights working>	0.0	0.0
<no starter motor turning>	0.0	0.0
<car doesn't start>	1.0	1.0

  Then, the parsing of the syntax rules will give the tree of Figure 7, together with the possibility and necessity values noted there. All the other alternatives of the "car out of order" rule were rejected.

Example 2: Full adder fault diagnostic procedure

  The second example has been taken form {27}, but it will be solved using the extensions described in section 6.8 and mainly the reduction operation and the contrapositives of a logic rule. To diagnose faults in a device we need information like:

  – Theoretical data: rules proving the correctness of the device.

  – Achievable data: propositions describing specific status of components of the device that can be achieved by a person.

  – Observable data: propositions describing specific status of components of the device that can be achieved by a person.

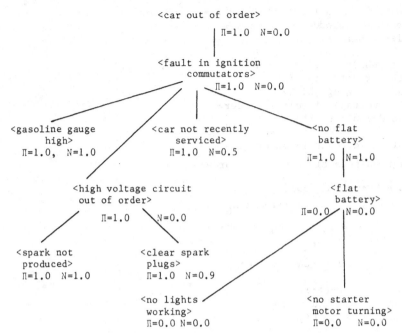

Fig. 7 Parse tree of example 1.

The device of our example is the full adder. A full adder is a very common digital circuit, a 1-bit adder with carry in and carry out, used in sets of n elements to provide a n-bit adder. The graphical representation of the design of the full adder is shown in Figure 8. Its circuit consists of two XOR gates, two AND gates and one OR gate. The full adder has three inputs which are the two bits to be added and the carry in. It also has two outputs which are the result of the addition and the carry out. Not only the gates but the connections among the gates, the inputs and the outputs can also be regarded as components of the full adder. The truth table of the full adder, and the truth tables of the gates and the connections are represented in Table III.

In this example the theoretical data are the rules which describe the behavior of the gates, the connections and the full adder. They correspond to the rules (39) through (79) of the knowledge base, where the contrapositives of the rules are also included. The achievable data are the propositions that describe the input data of the full adder. This means that the inputs of the full adder were set to some values (true or false) by a person. They correspond to the facts (2) through (7) of the knowledge base. The observable data are the prorositions that describe the output data of the full adder, observed by the person which was testing the device. They correspond to the facts (8), (9), (10) of the knowledge base.

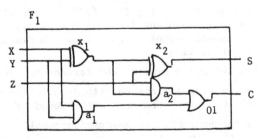

Fig. 8 Graphical representation of the full adder circuit.

Table 3

Truth tables of the gates											Truth table of the full-adder				

Table 3

Truth tables of the gates

AND-gates		OR-GATES			XOR-gates			Connections		
0	1	0	0	0	0	0	0	0	0	0
0	1	0	0	1	1	1	0	1	1	1
1	0	0	1	0	1	0	1	1		
1	1	1	1	1	1	1	1	0		

Truth table of the full-adder

X	Y	X	S	C
0	0	0	0	0
0	0	1	1	0
0	1	0	1	0
0	1	1	0	1
1	0	0	1	0
1	0	1	0	1
1	1	0	0	1
1	1	1	1	1

The notation used in the following example is: The arguments with lower case letters are constants while arguments with upper case letters are variables.

i1,i2,i3 are inputs of the full adder, or of the gates.
o1,o2   are outputs of the full adder, or of the gates.
x1,x2   are the XOR gates.
a1,a2   are the AND gates.
or is   the or gate.
f1 is   the full adder.
t1,t2   are the 1st time period and the 2nd time period accordingly.

value(X,Y,Z,W) The value of input/output X of the element Y at a time Z is W.
conn(X,Y,Z,W) The input/output X of the element Y is connected to the input/output Z of the element W.
xorg(X)   The element X is an xor gate.
org(X)   The element X is an or gate.
andg(X)   The element X is an and gate.

The entries of the knowledge base start from 2) because the first entry is occupied by the top goal to be proved.

2.1. Achievable data

2) value(i1,f1,t1,true).
3) value(i2,f1,t1,false).
4) value(i3,f1,t1,false).
5) value(i2,f1,t2,false).
6) value(i1,f1,t2,true).
7) value(i3,f1,t2,true).

2.2. Observable data

8) value (o1,f1,t1,false).
9) value (o2,f1,t1,false).
10) value (o2,f1,t2,false).

2.3. Assumptions

The rules that describe the behavior of the components of the full adder, are not enough for the theorem prover to find all the faulty components. There is also need for some assumptions to be made about the malfunctioning device. The assumptions made at this example are:

    a. The connections of the circuits are correct.
    b. There is at most one faulty component in the device.

2.3.1. Connections always correct assumption

```
11)  conn(il,fl,il,xl).
12)  conn(il,fl,il,al).
13)  conn(i2,fl,i2,xl).
14)  conn(i2,fl,il,al).
15)  conn(i3,fl,i2,x2).
16)  conn(i3,fl,il,a2).
17)  conn(ol,xl,il,x2).
18)  conn(ol,xl,i2,a2).
19)  conn(ol,al,i2,or).
20)  conn(ol,a2,il,or).
21)  conn(ol,x2,ol,fl).
22)  conn(ol,or,o2,fl).
23)  conn(il,xl,il,fl).
24)  conn(il,al,il,fl).
25)  conn(i2,xl,i2,fl).
26)  conn(il,al,i2,fl).
27)  conn(i2,x2,i3,fl).
28)  conn(il,a2,i3,fl).
29)  conn(il,x2,ol,xl).
30)  conn(i2,a2,ol,xl).
31)  conn(i2,or,ol,al).
32)  conn(il,or,ol,a2).
33)  conn(ol,fl,ol,x2).
34)  conn(o2,fl,ol,or).
```

2.3.2. Single fault assuption (SFA)

If we ask the theorem prover about the malfunction of the XOR gate x1, we must assume that the rest of the gates function properly. Thus, we include the following facts in the knowledge base.

```
35)  xorg(x2).
36)  andg(al).
37)  andg(a2).
38)  org(or).
```

2.4. XOR gate behavior

```
39)  value(ol,D,T,true)=xorg(D),value(il,D,T,true),value(i2,D,T,false).
40)  ∿xorg(D)=∿value(ol,D,T,true),value(il,D,T,true),value(i2,D,T,false).
41)  ∿value(il,D,T,true)=xorg(D),∿value(ol,D,T,true),value(i2,D,T,false).
42)  ∿value(i2,D,T,false)=xorg(D),value(il,D,T,true),∨value(ol,D,T,true).
43)  value(ol,D,T,false)=xorg(D),value(il,D,T,true),value(i2,D,T,true).
44)  ∿xorg(D)=∿value(ol,D,T,false),value(il,D,T,true),value(i2,D,T,true).
45)  ∿value(il,D,T,true)=xorg(D),∿value(ol,D,T,false),value(i2,D,T,true).
46)  ∿value(i2,D,T,true)=xorg(D),value(il,D,T,true),∨value(ol,D,T,false).
47)  value(ol,D,T,true)=xorg(D),value(il,D,T,false),value(i2,D,T,true).
48)  ∿xorg(D)=∿value(ol,D,T,true),value(il,D,T,false),value(i2,D,T,true).
49)  ∿value(il,D,T,false)=xorg(D),∿value(ol,D,T,true),value(i2,D,T,true).
50)  ∨value(i2,D,T,true)=xorg(D),value(il,D,T,false),∿value(ol,D,T,true).
51)  value(ol,D,T,false)=xorg(D),value(il,D,T,false),value(i2,D,T,false).
52)  ∿xorg(D)=∿value(ol,D,T,false),value(il,D,T,false),value(i2,D,T,false).
53)  ∿value(il,D,T,false)=xorg(D),∨value(ol,D,T,false),value(i2,D,T,false).
45)  ∨value(i2,D,T,false)=xorg(D),value(il,D,T,false),∨value(ol,D,T,false).
```

2.5. AND gate behavior

55) value(o1,D,T,true)=andg(D),value(i1,D,T,true),value(i2,D,T,true).
56) ᴠandg(D)=ᴠvalue(o1,D,T,true),value(i1,D,T,true),value(i2,D,T,true).
57) ᴠvalue(i1,D,T,true)=andg(D),ᴧwalue(o1,D,T,true),value(i2,D,T,true).
58) ᴧwalue(i2,D,T,true)=andg(D),value(i1,D,T,true),ᴧwalue(o1,D,T,true).
59) value(o1,D,T,false)=andg(D),value(i1,D,T,false),
60) ᴧandg(D)=ᴧwalue(o1,D,T,false),value(i1,D,T,false).
61) ᴧwalue(i1,D,T,false)=andg(D),ᴧwalue(o1,D,T,false).
62) value(o1,D,T,false)=andg(D),value(i2,D,T,false).
63) ᴧandg(D)=ᴧwalue(o1,D,T,false),value(i2,D,T,false).
64) ᴠvalue(i2,D,T,false)=andg(D),ᴧwalue(o1,D,T,false).

2.6. OR gate behavior

65) value(o1,D,T,false)=org(D),value(i1,D,T,false),value(i2,D,T,false).
66) ᴠorg(D)=ᴧwalue(o1,D,T,false),value(i1,D,T,false),value(i2,D,T,false).
67) ᴧwalue(i1,D,T,false)=org(D),ᴠvalue(o1,D,T,false),value(i2,D,T,false).
68) ᴠvalue(i2,D,T,false)=org(D),value(i1,D,T,false),ᴧwalue(o1,D,T,false).
69) value(o1,D,T,true)=org(D),value(i1,D,T,true).
70) ᴠorg(D)=ᴧwalue(o1,D,T,true),value(i1,D,T,true).
71) ᴧwalue(i1,D,T,true)=org(D),ᴠvalue(o1,D,T,true).
72) value(o1,D,T,true)=org(D),value(i2,D,T,true).
73) ᴧorg(D)=ᴠvalue(o1,D,T,true),value(i2,D,T,true).
74) ᴧwalue(i2,D,T,true)=org(D),ᴠvalue(o1,D,T,true).

2.7. Connections behavior

75) value(O1,D1,T,V)=conn(O2,D2,O1,D1),value(O2,D2,T,V).
76) ᴧconn(O2,D2,O1,D1)=ᴧwalue(O1,D1,T,V),value(O2,D2,T,V).
77) ᴧwalue(O2,D2,T,V)=conn(O2,D2,O1,D1),ᴧwalue(O1,D1,T,V).

2.8. Rule about values

78) ᴧwalue(X,Y,Z,true)=value(X,Y,Z,false).
79) ᴧwalue(X,Y,Z,false)=value(X,Y,Z,true).

The above knowledge base is not the initial one, but contains additional rules which correspond to the general contrapositives of other rules. All such rules are gathered together, in small groups. It must be noted that the symbol "ᴠ" corresponds to the negation symbol. Thus the proposition ᴠxorg(x1) is the negation of the proposition xorg(x1). For example rules (52), (53), (54) are the general contrapositives of rules (51). The general contrapositives were not initially included in the knowledge base, but have been automatically generated by a preprocessing procedure.

The rules have been purposely put in such a sequence so that the deduction become simple. That's how we have resulted in a small number of inference steps. However, if we alternate the sequence of the rules in the knowledge base, another solution path will be found when the Goal will be proved. In the later case, the system uses the ME reduction rule and a left-recursion-detection rule. The left-recursion-detection rule has been added in order to avoid infinite loops. The inference steps that prove the goal ᴠxorg(x1) are presented in the following lines.

Theorem to be proved:

1.   ᴠxorg(x1)
    40) ᴠxorg(D)=ᴧwalue(o1,D,T,true),value(i1,D,T,true),value(i2,D,T,false).
    x1/D

2. $\sim$value(01,x1,T,true)
    77) $\sim$value(02,D2,T,V)=conn(02,D2,01,D1),$\sim$value(01,D1,T,V).
        01/02,x1/D2,true/V

3.   conn(o1,x1,01,D1)
    17) conn(01,x1,i1,x2)
        PROVED for i1/01,x2/D1

4. $\sim$value(i1,x2,T,true)
    41) $\sim$value(i1,D,T,true)=xorg(D),$\sim$value(o1,D,T,true),value(i2,D,T,false).
        x2/D

5.   xorg(x2)
    35) xorg(x2) PROVED

6. $\sim$value(o1,x2,T,true)
    77) $\sim$value(02,D2,T,V)=conn(02,D2,01,D1),$\sim$value(01,D1,T,V).
        o1/02,x2/D2,true/V

7.   conn(01,x2,01,D1)
    21) conn(01,x2,01,f1)
        PROVED for 01/01,f1/D1

8. $\sim$value(01,f1,T,true)
    78) $\sim$value(X,Y,Z,true)=value(X,Y,Z,false)
        o1/X,f1/Y

9.   value(o1,f1,Z,false)
     8) value(o1,f1,t1,false)
        PROVED for t1/Z
        $\sim$value(01,f1,t1,true) of step 8 has been PROVED
        $\sim$value(01,x2,t1,true) of step 6 has been PROVED

10.   value(i2,x2,t1,false) of step 4 to be proved
    75) value(01,D1,T,V)=conn(02,D2,01,D1),value(02,D2,T,V)
        i2/01, x2/D1, t1/T, false/V

11.   conn(02,D2,i2,x2)
    15) conn(i3,f1,i2,x2)
        PROVED for i3/02, f1/D2

12.   value(i3,f1,t1,false)
     4) value(i3,f1,t1,false) PROVED
        value(i2,x2,t1,false) of step 10 has been PROVED
        $\sim$value(i1,x2,t1,true) of step 4 has been PROVED
        $\sim$value(o1,x1,t1,true) of step 2 has been PROVED

13.   value(i1,x1,t1,true)
    75) value(01,D1,T,V)=conn(02,D2,01,D1),value(02,D2,T,V)
        i1/01, x1/D1, t1/T, true/V

14.   conn(02,D2,i1,x1)
    11) conn(i1,f1,i1,x1)
        PROVED for i1/02, f1/D2

15.   value(i1,f1,t1,true)
     2) value(i1,f1,t1,true) PROVED
        value(i1,x1,t1,true) of step 13 has been PROVED

16.   value(i2,x1,t1,false)
    75) value(01,D1,T,V)=conn(02,D2,01,D1),value(02,D2,T,V)
        i2/01, x1/D1, t1/T, false/V

17.   conn(02,D2,i2,x1)
    13) conn(i2,f1,i2,x1)
        PROVED for i2/02, f1/D2

18.   value(i2,f1,t1,false)
    3) value(i2,f1,t1,false) PROVED
       value(i2,x1,t1,false) of step 16 has been PROVED
       Hence the top goal ∿xorg(x1) has also been proved.

    If we include the fact xorg(x1) instead of the fact xorg(x2) then
we may ask about the malfunction of x2 XOR gate. The top goal will now
become ∿xorg(x2) and the theorem prover will again answer positively. However the theorem prover will not give an answer to the questions
∿andg(a1), ∿andg(a2), ∿org(or). We may conclude that under the above
assumptions the only suspect components for malfunction are the x1, x2
gates.

8. CONCLUSION

    Actually no single methodology exists for the construction of expert
systems for decision making and process fault diagnosis. Several efficient designs have been proposed for fault diagnosis in particular domains
such as medical, electonic, computer H/W and S/W, rotating machines, and
industrial process diagnosis. Very broadly, these approaches fall in two
classes, namely: (i) shallow reasoning approach, and (ii) deep knowledge
approach.

    The attribute grammar approach presented in this chapter offers a
convenient frameworks, for building and using expert systems, which combines both declarative (factual) and inferential (procedural) knowledge
in a single tool. Since presently many implementations of compilers and
interpreters for attribute grammars'processing exists, this approach is
very promising for treating practical problems of intelligent decision
making, diagnosis and supervisory control. The authors are currently working in both direction of furhter developing the attribute grammar tool
and using it in engineering problems.

REFERENCES

1. A.V. Aho and J.D. Ullman, *The Theory of Parsing, Translation and
   Compiling*, Prentice-Hall, Englewood Cliffs, NJ (1972).

2. D.E. Knuth, Semantics of Context-Free Languages, *Mathem. Syst. Theory*,
   2:127-145 (1968).

3. W.M. Waite and G. Goos, *Compiler Construction*, Springer-Verlag (1984).

4. D.S. Nau, Expert Computer Systems, *Computer*, 16 (2):63-85 (1983).

5. G. Papakonstantinou and J. Kontos, Knowledge Pepresentation with
   Attribute Grammars, *The Computer Journal*, 29 (3):241-245 (1986).

6. G. Papakonstantinou, C. Moraitis and T. Panayiotopoulos, An Attribute
   Grammar Interpreter as a Knowledge Engineering Tool, *Angewandte
   Informatik*, 9/86:382-388 (1986).

7. P. Deransart and Maluszynski, Relating Logic Programs and Attribute
   Grammars, *J. Logic Programming*, 2:119-155 (1985).

8. B. Arbab, Compiling Circular Attribute Grammars into Prolog, *IBM J,
   Res. Devel.* 30 (3):294-309 (1986).

9. G. Papakonstantinou, An Interpreter of Attribute Grammaras and its Application to Waveform Analysis, *IEEE Trans. Software Eng.*, SE-7: 279-283 (1981).

10. K.J. Räihä, Bibliography on Attribute Grammars, *SIGPLAN Notices*, 15 (5):35-44 (1980).

11. G.V. Bochmann, Semantic Evaluation from Left to Right, *CACM*, 19 (2): 55-62 (1976).

12. R.W. Floyd, The Syntax of Programming Languages: A Surrey, *IEEE Trans. Electron. Comp.*, AC 13 (4):346-353 (1964).

13. W.H. Tsai and K.S. Fu, Attribute Grammars, A Tool for Combining Syntactic and Statistical Approaches to Pattern Recognition, *IEEE Trans. SMC*, SMC-10 (12):873-885 (Dec., 1980).

14. J. Earley, An efficient Context-Free Parsing Algorithm, *CACM*, 13(2): 94-102 (1970).

15. E. Shortliffe, *Computer-Based Medical Consultations: MYCIN*, Elsevier, New York (1976).

16. A.P. Dempster, Upper and Lower Probabilities Induced by a Multivalued Mapping, *Annals Mathem. Statistics*, 38:325-339 (1967).

17. D. Dubois, and H. Prade, *Théorie des Possibilités*, Masson, Paris (1985).

18. H. Prade, A Computational Approach to Approximate and Plausible Reasoning with Applications to Expert Systems, *IEEE Trans. PAMI*, PAMI-7 (3):260-283 (1985).

19. C. Moraitis, G. Papakonstantinou and S. Tzafestas, Attribute Grammars as a Diagnostic Tool, in *System Fault Diagnostics, Reliability and Related Knowledge-Based Approaches*, (Edited by: S. Tzafestas, M. Singh and G. Schmidt) Vol. 2:53-62, D. Reidel, Dordrecht (1987).

20. C. Moraitis, Approximate and Plausible Reasoning Using Attribute Grammars, *Research Report Computer Div.*, Natl. Tech. Univ. of Athens, Athens (1986).

21. M. Cayrol, H. Farreny and H. Prade, Fuzzy Pattern Matching, *Kybernetes*, 11:103-116, (1982).

22. M.E. Stickel, A Prolog Technology Theorem Prover, *Proc. Intl. Symp. on Logic Programming*, Atlantic City, New Jersey, 211-213 (1984).

23. Z.D. Umrigar and V. Pitchumani, An Experiment in Programming with Full First-Order Logic, *Proc. Intl. Symp. on Logic Programming*, Boston, MA (1985).

24. W. Bibel and Ph. Jorrand (eds.), *Fundamentals of Artificial Intelligence*, LNCS-232, Springer-Verlag (1986).

25. T. Panayiotopoulos, G. Papakonstantinou and G. Stamatopoulos, An Attrinute Grammar-Based Theorem Prover, *Research Report, Computer Eng. Div.*, Natl. Tech. Univ. of Athens, Athens (1987).

26. T. Panayiotopoulos, G. Papakonstantinou and G. Stamatopoulos, A Theorem Prover as a Fault Diagnostic Tool, in *System Fault Diagnostics,*

Reliability and Related Knowledge-Based Approaches (Edited by: S. Tzafestas, M. Singh and G. Schmidt) Vol. 2:43-52, D. Reidel, Dordrecht (1987).

27. M.R. Genesereth, The Use of Design Descriptions in Automated Diagnosis, *Artificial Intelligence,* 24:411-436 (1984).

28. S.G. Tzafestas, Knowledge-Based Approach to System Modelling, Diagnosis, Supervision and Control, *Proc. IFAC/IMACS Symp. on Simulation of Control Systems* (Edited by I. Troch, P. Kopacek and F. Breitenecker) 17-31 (Sept., 1986).

29. S.G. Tzafestas, Artificial Intelligence Techniques in Control, *Proc. IMACS Symp. on AI, Expert Systems and Languages in Modelling and Simul.* (Edited by C. Kulikowski and G. Ferrate) 55-67 (June, 1987).

30. S.G. Tzafestas, Ststem Fault Diagnosis Using the Knowledge-Based Methodology, in *Fault Diagnosis in Dynamic Systems: Theory and Applications* (Edited by R. Patton, P. Frank and R. Clark) Ch. 15, *Prentice Hall Intl. (UK) Ltd.* (1988).

31. S.G. Tzafestas, A Look at the Knowledge-Based Approach to System Fault Diagnosis and Supervisory Control, in *System Fault Diagnostics, Reliability and Related Knowledge-Based Approaches* (Edited by S. Tzafestas, M. Singh and G. Schmidt) Vol. 2:3-15, D. Reidel, Dordrecht (1987).

32. T. Panayiotopoulos, G. Papakonstantinou and G. Stamatopoulos, Attribute Grammars and Logic Programming, *Angewandte Informatik* (AI-debot paper, to appear).

33. T. Panayiotopoulos, G. Papakonstantinou and G. Stamatopoulos, An Attribute Grammar Based Theorem Prover, *Information and Software Technology* (provisionally accepted).

*Paper originally published in issue 4:87 of Journal A.

8

COUPLING SYMBOLIC AND NUMERICAL COMPUTATION FOR INTELLIGENT SIMULATION

Guy O. Beale* and Kazuhiko Kawamura**

*Department of Electrical and Computer Engineering
George Mason University, Fairfax, Virginia, USA

**Center for Intelligent Systems, Vanderbilt University
Nashville, Tennessee, USA

1. INTRODUCTION AND BACKGROUND

For the past several years, expert systems technology has been gaining in popularity and acceptance in the engineering community (1). Applications range from diagnostic and repair domains to design assistance (2,3). However, common knowledge-based expert systems are often not well suited to dealing with numerical processing due to their focus on symbolic reasoning. In contrast, numerical techniques, required for most engineering applications, can provide rapid high precision solutions to mathematically-formulated problems, but they often cannot provide insight into the problem solving process or interpretation of the results. Therefore, since there is a need for results generation and results interpretation in complex problems, there is a need to couple symbolic and numerical techniques. Such coupled systems promise to integrate the explanation and problem solving capabilities of knowledge-based systems with the precision of traditional numerical computing.

One application area where this coupling of symbolic and numerical computing will have great benefit is the digital simulation of complex dynamic systems. This chapter describes the design principle and architecture of one such coupled system, named NESS (NASA Expert Simulation System) (4). NESS assists the user in running digital simulations of dynamic systems; interprets the output data to determine system characteristics; and, if the output does not meet the performance specifications, recommends a suitable series compensator to be added to the simulation model.

In 1985, a workshop was sponsored by the American Association for Artificial Intelligence to discuss issues involving coupling symbolic and numeric computing techniques (5). In the various papers presented at this workshop, two major reasons for coupling symbolic and numerical processing emerged (6). The first reason is to aid users of complex numerical algorithms and programs. Oftentimes, insight is required in order to: (1) obtain a solution at all, (2) learn to utilize the tools at the user's disposal, or (3) interpret computed results. As previously mentioned, this insight generally has not been provided by traditional numerical computing algorithms. Coupled systems promise to integrate this insight with the computational power of numerical methods.

The second major reason for coupling symbolic and numerical computing is to handle problems involving ambiguous, contradictory, and imprecise data (6). In large complex systems, there will be parts which are not sufficiently well understood to be amenable to numerical techniques. A more robust problem solving environment is needed, and it is felt that coupling symbolic and numerical processing can provide that power and robustness.

The 1987 IEEE International Symposium on Intelligent Control also addressed the issue of coupled systems. Vachtsevanos and Davey (8) describe an expert system to perform on-line fault diagnosis in a thermal control system. The intelligent system uses fuzzy logic to locate and isolate the fault. Villa (9) presents an application of hybrid symbolic and numerical computing in the design of control systems characterized by complexity and uncertainty. The conclusion is that the attributes of the knowledge-based and numerical systems can complement each other.

2. NESS DESIGN PRINCIPLE

2.1. Overview

The hypothesis for using a knowledge-based system in conjunction with numerical methods for the digital simulation of a dynamic system is that it takes specialized knowledge to obtain a stable and accurate simulation of a complex system. Choosing a simulation method and the simulation time step are often critical choices which must be made. Once an acceptable simulation of the given system is achieved, the performance of the system itself must be evaluated relative to its goals. If that performance is not satisfactory, then modification of the system through the addition of compensating elements might be required. The knowledge necessary to simulate the system accurately, interpret the performance in terms of goals, and recommend compensators might not be directly available to the user. An expert in the domain of the system to be simulated may not have the necessary detailed knowledge in numerical methods and control system synthesis.

There is one feature which makes this particular application different from the usual simulation problem. No explicit information on the dynamics of the system being simulated is given to the knowledge-based system running the simulation. The only knowledge that can be used during execution of the simulation and design of the compensator comes from input/output measurements made during the simulation. The knowledge-based system is therefore dealing with a "black box" during the simulation, and the results of analyses and compensator designs will reflect this. This class of coupled systems is called loosely coupled or shallow coupled systems (7).

2.2. Functions of NESS

The design principle guiding the development of NESS is that there will be a clear separation between the simulation model and the expert system. The software code modeling the system to be simulated, along with a glossary of state variables, parameter values, and time constraints, would be generated by the user. NESS would then use its knowledge to: (1) determine the method of numerical integration, (2) choose the value of the simulation time step, (3) initiate the execution of the simulation program, (4) present and interpret the results, and (5) recommend any compensators required to achieve the desired performance. If the user wanted to evaluate the modified system, the modifications would be incorporated into the code for the system model, and the simulation would be repeated with the new code. With the intelligent simulation, the user can be confident of an accurate simulation of the given model and is provided with recommendations on how to improve the system performance. Therefore, the time required for the iterative design and verification process should be lessened.

Figure 1 is a block diagram which illustrates this design principle. The simulation model is the software code implementing the equations which mathematically model the actual system. This code is generated from the user's description of the system. Also created at this time is the Glossary. The Glossary is a knowledge base about the system which specifies: (1) the state variables in the system, (2) the initial values for the state variables, (3) the values for system parameters, and (4) any constraints on the size of the time steps used in the simulation. NESS reads this knowledge base, creates an interface with the simulation model, and instantiates the model with the values from the knowledge base. In general, this interface will include the numerical integration algorithms for all of the state variables. The user interactively inputs the system performance specifications directly into NESS. Once the model is instantiated, the simulation may be performed under the direction of NESS in either an automatic or a manual mode. In the automatic mode, NESS performs as many actions as possible without further user interaction, in-

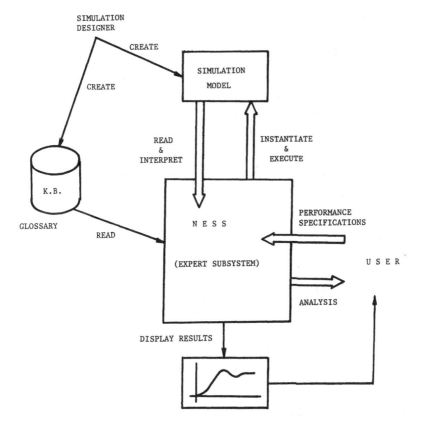

Fig. 1 NESS design principle.

cluding compensator design. In the manual mode, the user is given the option at each step either of agreeing with a suggestion by NESS or of providing direction to NESS.

Simulation of the system can be done in either the time domain or the frequency domain. The results of the simulation are interpreted by NESS realtive to the performance specifications supplied by the user. For example, the transient performance might be described by the percent overshoot in the time domain or the phase margin in the frequency domain. Steady-state accuracy to step or ramp inputs can be obtained directly in the time dommin or inferred from frequency domain data. Based on the performance analysis, the user would be told whether or not all specifications are met. If they are not, a lead, lag, or lag-lead compensator would be recommended, depending on the type of performance error.

3. SYSTEM ARCHITECTURE

NESS consists of two subsystems, the Expert Subsystem and the Numeric Subsystem. Figure 2 illustrates this architecture. The two subsystems are described in the following paragraphs.

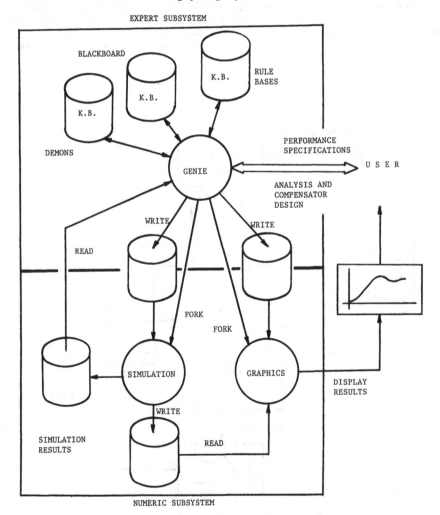

Fig. 2 NESS system architecture.

3.1. Expert Subsystem

The Expert Subsystem contains the knowledge required to assist a user in running a simulation model and to display and interpret the simulation results. If the performance of the model is not satisfactory, a series compensator will be recommended that could be added to achieve the desired results.

The Expert Subsystem is designed in a modular fashion. It contains six function-specific modules, all of which are under the control of a seventh module called the Planner. The architecture of this Subsystem is illustrated in Figure 3. The Expert Subsystem is written in FRANZ LISP (10) and uses an inference engine named GENIE (GENeral Inference Engine), developed at Vanderbilt University (11).

Each function-specific module contains rules and demons (special-purpose LISP functions) needed to implement a particular function. These modules use the blackboard to post decisions made by themselves, and these decisions are available for other modules to act upon. All modules share the data on the blackboard, but the rules and demons are specific to the particular module. Hence changes can be made to the internal structure of a module without affecting the internal structure of others. More function-specific modules can easily be added to expand the scope of NESS, if needed. The following paragraphs provide a brief description of each module in the Expert Subsystem.

3.1.1. Planner

This is the controller of the Expert Subsystem. The Planner is the only module which communicates with other modules. Various strategies can be programmed in the Planner to provide an orderly sequencing of the functions in NESS. The Planner carries out the operations of NESS by calling appropriate modules. For example, it calls the Simulation Executor to run the simulation program and, when required, the Graphics module to

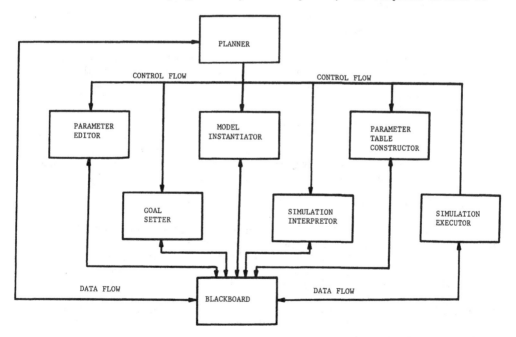

Fig. 3 Expert Subsystem architecture.

display results. The Planner is also able to look at the decisions posted
by other modules to determine future strategies or methods of operation.

3.1.2. Goal Setter

This module allows the user to set the goals for the simulation in terms
of the response type: step, ramp or frequency. He can also enter the desired
performance standards for the simulation. These include gain and phase margins
for the frequency response: rise time, settling time, and steady-state error
for the ramp response.

3.1.3. Parameter Table Constructor

This module reads the Glossary and adds the knowledge found there
to the blackboard. A typical glossary contains a list of all state variables and model coefficients present in the simulation model, along with
their attributes. The contents of the Glossary are transformed into frames which are the basic data structure available in GENIE. One frame is
created for each state variable and model coefficient. A typical frame is:

```
(sv1
     (name (ThetaX))
     (description     (Angular position about the X-axis))
     (default_value  (0.01))
     (units      (radians))
     (deltat_type    (v))
     (delta_def_val    (n))
     (integration    (n))
     (module    (main)))
```

Here "sv1" denotes the first state variable in the Glossary. Its
attributes are name, description, default_value, units, deltat_type (variable or fixed), deltat_def_val (default value for time step), integration, and module. All attributes have values which can be found in the
Glossary and in the frame.

3.1.4. Model Instantiator

This module is responsible for obtaining initial values for all parameters required by the simulation program. They include initial conditions for the state variables, values for the model coefficients, desired
input and output points, the methods of integration, the value for the
rate of integration, and the parameter values required for step, ramp, or
frequency response. The user can review the attributes of any parameter
and the input and output points. The method of integration is initially
set to Euler for all state variables. The user can select a different
mtthod of integration for subsequent simulations. The rate of integration
is set to the value calculated by the Parameter Table Constructor.

3.1.5. Simulation Executor

This module first writes all the parameter values obtained via user
interaction to the data files. Then it forks a task to run the simulation
program through the FRANZ LISP function "*process". The current LISP task
is suspended and the task forked by "*process" begins to execute. After
the simulation program completes execution, the forked task dies and
control is returned to the original LISP task. At this stage it can be
determined whether the simulation program executed successfully or not.
This knowledge is posted on the blackboard for the Planner and the Simulation Interpretor.

3.1.6. Simulation Interpretor

This module analyzes the simulation results and categorizes them to be one of three types: error-prone, not-yet-reached steady-state, or stable. An error-prone simulation is one in which a Fortran run-time error occurred during execution. A not-yet-reached-steady-state simulation is one which failed to reach steady state during the allowed simulation time. A stable simulation is one which reached steady-state conditions. If the simulation has reached steady state, the simulation interpretor reads the results generated. If it is decided that a compensator is needed to enhance the performance of the simulation, this module recommends a suitable compensator.

3.1.7. Parameter Editor

This module allows the user to edit parameter values. The user is presented with a menu of options as follows:

1) Exit
2) State Variables
3) Model Coefficients
4) Integration Method
5) Input/Output Points
6) Time/Frequency Parameter(s)

Using the Editor, the user is able to change any of the parameter values associated with the simulation. Following these changes, the simulation may be executed again.

3.2. Numerical Subsystem

Figure 4 illustrates the Numeric Subsystem architecture. The Numeric Subsystem is composed of four modules which are described below.

3.2.1. Main

Figure 5 illustrates the flowchart of the main program. This module is responsible for calling other modules. The interface between the Simulation Model, the Utilities Interface, and the Main modules is built around three FORTRAN "COMMON" areas.

3.2.2. Simulation Model

This is the "coded model" module provided by the simulation designer. It is the application code for the dynamic system being simulated. The top-level structure of this module communicates with the remaining modules through the COMMON area in Main. Therefore, information regarding state variables and model parameters is passed through the state variable (SV) and model coefficient (MC) data structures, respectively.

3.2.3. Utilities Interface

3.2.3a. Integrator Submodule. The Integrator Submodule, one of two submodules in the Utilities Interface, performs several functions. At the beginning of the simulation it computes all of the initial values required by the integration operators. During the simulation, the submodule numerically integrates the state variables at each time step. The Integrator is able to distinguish true state variables, needing integration, from dummy state variables (i.e., input points and access points). The

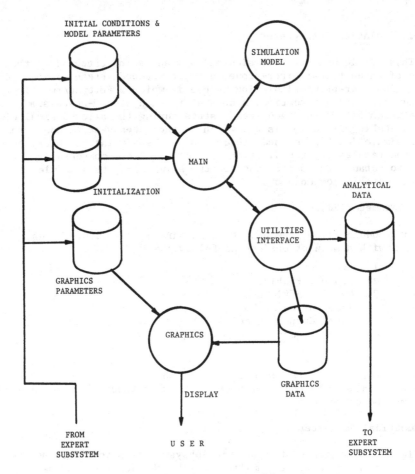

INITIAL CONDITIONS &
MODEL PARAMETERS

SIMULATION
MODEL

MAIN

INITIALIZATION

ANALYTICAL
DATA

UTILITIES
INTERFACE

GRAPHICS
PARAMETERS

GRAPHICS

GRAPHICS
DATA

DISPLAY

FROM
EXPERT
SUBSYSTEM

U S E R

TO
EXPERT
SUBSYSTEM

Fig. 4 Numeric subsystem architecture.

Integrator submodule also updates the actual input signal at each step
and applies it to the appropriate signal point.

   3.2.3b. Analysis Submodule. The Analysis submodule is called by the
Integrator submodule to process the data following the computation of each
state variable. The Analysis submodule is capable of performing time-
domain and frequency-domain analyses. In time-domain analysis two types
of inputs, step and ramp, are provided; transient and steady-state performance of the system can be determined. In frequency-domain analysis, a
pseudo open-loop frequency response is analyzed to determine the system
stability margins. The strategies that apply to each type of input are:

1) For a step input, the Analysis submodule will monitor and record the
    output to compute system characteristics such as delay time, rise
    time, peak value, peak time, settling time, and steady-state error.

2) For a ramp input, the slopes at input and output will be compared at
    each time interval. From these, the steady-state condition can be de-
    tected and used to calculate the steady-state error. If the system is
    less than type 3, the system type can be determined from the relation-
    ship between output slope and input slope and the value of steady-
    state-error.

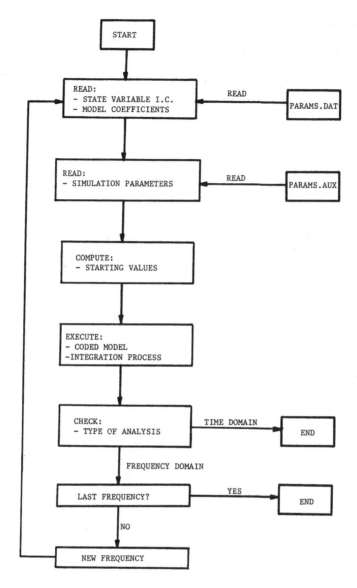

Fig.5 NESS main program flow chart.

3) For a sinusoidal input, the input will be injected into a prespeci-
    fied access point once steady state is reached. By performing the
    pseudo open-loop frequency analysis, a Bode plot will be generated,
    and the values of gain mmrgin and phase margin can be obtained.

3.2.4. Graphics

    This module uses the TCS Tektronix graphics package to display the
simulation results. Plots in both the time and frequency domain can be
provided. This module is activated by the Planner each time the user
requests a graphics display.

3.3. Subsystem Coupling

The expert subsystem uses the FRANZ LISP "*process" function to fork a descendent process which runs the executable simulation program in the numeric subsystem. Before forking the descendent process, the expert subsystem creates two files containing the simulation model parameter values. When the simulation run is completed, the numeric subsystem writes the results to a file which in turn is read by the expert subsystem.

4. NESS KNOWLEDGE BASE

4.1. Overview

The knowledge incorporated into NESS is in two areas. The first area is numerical methods as related to the simulation of systems represented by ordinary differential equations. This knowledge provides the ability to choose a numerical integration method and a time increment in order to produce an accurate and stable simulation of the system. The choice of method and step size might have to be iterative since very little knowledge about the system's eigenvalues is known to NESS; the simulation results provide most of the knowledge about the system. Due to the interaction between NESS and the system model during simulation, only linear multistep integration algorithms, which require one calculation of the differential equations per time step, are used (12,13).

The other area of knowledge relates to interpretation of the simulation results. The first task of interpretation for NESS is to determine whether or not the results accurately represent the performance of the actual system. For example, if the simulation appears to be unstable, the cause could be an improper choice of integration method or step size, or it could be a very accurate simulation of an inherently unstable process. This needs to be decided before any further interpretation can be done. Once an acceptable simulation of the system is obtained, the results need to be analyzed from the standpoint of performance. Properties of both the transient response and steady-state accuracy can be specified by the user. NESS has the intelligence to analyze the simulation results, infer the response characteristics from the results, and recommend a compensation technique if the performance is not satisfactory. Classical techniques for the design of compensators using Bode plots and root locus plots are part of the knowledge base in NESS. Lag compensators are designed when the transient performance is satisfactory but the steady-state accuracy needs to be improved. Lead compensators are designed when the accuracy is acceptable but the transient performance needs improvement. Lag-lead compensators are designed when both accuracy and transient response must be improved.

4.2. Model Instantiation

The intelligence associated with the Model Instantiator relates to: (1) the instantiation of the model parameters and (2) the selection of the integration method and the step size required to ensure stability. A brief description of the rules in the current version of the Model Instantiator follows.

4.2.1. Instantiation of Model Parameters

The Model Instantiator is activated by the Planner to perform all necessary instantiations of the simulation model parameters. The instantiation of state variables and model coefficients is carried out separately by the Planner and is implemented as two independent procedures, mi_sv and mi_mc. Both procedures make use of the same general rule base, mi_rb. The mi_rb rule base is composed of two rules, one for instantiating the state variables and one for instantiating the model coefficients. Each of these rules is run in backward chaining by mi_sv amd mi_mc, respectively. During their execution, mi_rb will prompt the user with a menu. Thus, the user is allowed to either set all parameters to default values, set some of them interactively, or set none of them to default. Once the user makes a choice, mi_rb will transfer the control to the rule bases mi_sv_rb and mi_mc_rb, which are executed in forward chaining.

Once the instantiation of the state variables and model coefficients is successfully performed, the Planner will call procedures to instantiate the input and output points. The user will be prompted with a menu of all the input points as defined in the Glossary. Following this, a menu of all the possible output points will be presented. The process of instantiating the simulation model parameters ends when the user makes his choices.

4.2.2. Selection of Integration Method and Step Size

The integrator module assumes that a state space representation of the system is given, and that the solution to the differential equation.

$$dX/dt = F(X,K,t), \qquad X(0) \text{ given}$$

is obtained using linear multistep integration operators (12) of the form

$$X(n) = a1 \ X(n-1) + a2 \ X(n-2) + \ . \ . \ . \ +$$
$$T \left[b1 \ F(n-1) + b2 \ F(n-2) + \ . \ . \ . \ . \right]$$

X is the state vector, K is the vector of model coefficients, t is time, T is the simulation time step, and nT is an arbitrary point in time. The state vector includes not only state variables, but also those locations in the simulation model where external signals can be injected to perform either a time- or frequency-domain analysis.

There are several reasons why linear multistep integration methods have been incorporated into NESS. They are listed and briefly discussed in this section.

4.2.2a. Current Practice at NASA. NASA extensively uses the Euler method of numerical integration. The form of the Euler equation is identical to that of linear multistep methods, and the Euler method can be thought of as a "one-step" linear multistep method.

4.2.2b. Extension of Euler's Method. Linear multistep methods are a natural extension of Euler's method. One equation can be programmed for all axplicit linear multistep methods; only the coefficients in the equation need be changed for different operators.

4.2.2c. Increased Accuracy. Higher accuracy can be obtained from using multistep methods, such as the Adams-Bashforth operators, with only a slight increase in the amount of computational burden.

4.2.2d. Exhibition of Intelligence. The use of more sophisticated integration algorithms allows NESS to exhibit more intelligence as well as producing a more accurate simulation. The selection of the integration method involves decisions concerning the required accuracy, the computation time available, and the maximum time step allowable for a stable simulation.

4.2.3. Selection Rules

The rules associated with the selection of the most appropriate integration method and step sizes for each state variable have been established as follows. First, during the process of instantiating all the simulation parameters, general knowledge regarding numtrical stability is obtained from the Glossary, which sets a maximum value for the step size and selects the Euler integration operator. Therefore, this information gives NESS a starting point to estimate the largest eigenvalue associated with the simulation model. The product of the largest eigenvalue and the simulation time step must fall within the region of absolute stability of the integration method for the simulation to be stable. Based on the estimate of the maximum eigenvalue and knowledge of the regions of absolute stability for the available integration methods, a limit can be put on the step size for any method chosen. Whenever a change is made to either integration method or step size, the requirements for absolute stability must be satisfied by NESS.

Therefore, when a simulation is found to be stable, and if the user asks for improvements in accuracy, NESS will prompt the user with a menu of options which contains an option to change integration operators. If a change of integration operators is required, the LISP procedure pe_integration2 is executed, and NESS gives the user the opportunity to select a new integration operator from a menu of options. The integration operators currently supposted by NESS are: Euler, 2nd order Adams-Bashforth, a three-step linear multistep method with an order of accuracy equal to 3, and the 4th order Adams-Bashforth method.

Two possible cases may occur during user-NESS interaction when selecting a new integration operator:

1) The user's new selected integration method has a higher order of accuracy than the one used for the initial simulation. In order to maintain a stable simulation, the LISP procedure chop_deltat reduces integration rate(s) by the power of 2 necessary to keep the same relationship between the time step and the boundary of absolute stability.

2) The user's new selected integration operator has a lower order of accuracy than the method originally used. In this case, another LISP procedure, high_deltat, increases the integration rate(s) by the appropriate power of 2.

4.3. Result Interpretation

The Simulation Interpreter classifies a particular simulation run into one of three types: "Error-Prone", "Not-Yet-Reached-Steady-State", and "Stable". After the results of the current run are presented to the user, the Simulation Interpreter rule base "strategy_rb" determines the next strategy to be used in order to improve the simulation results, if needed.

4.3.1. Error-Prone Simulation

An error-prone simulation is the one in which a Fortran run-time error has occurred during execution. This error can be due to various factors, such as an unstable simulation. This condition usually results in an overflow error during execution. In the manual mode, NESS asks the user whether the run-time error was caused by either overflow or underflow. If the user responds "YES", NESS reduces the rate of integration and reruns the simulation. If the response is "NO", NESS asks the user for further guidance.

In the autommtic mode, NESS reduces the rates of integration and reruns the simulation without user intervention. NESS will do this a maximum of two times. After two additional simulations with run-time errors, NESS goes from automatic to manual mode in order to get user input. An example of a rule which implementes this strategy follows:

IF [The user wants automatic mode]

and [a FORTRAN run-time error has occured]

THEN [Reduce integration rates by 4.0 and rerun simulation]

4.3.2. Not-Yet-Reached-Steady-State Simulation

A simulation may not reach steady state because: (1) the simulation was numerically unstable, (2) the simulation was run for a shorter time than needed to reach steady state, or (3) the simulation model is inherently unstable. NESS is not able to determine which of the above conditions caused the simulation to not attain a steady-state condition. In the manual mode it provides the user with options to rerun the simulation after reducing the rate of integration and/or increasing the simulation run time. The user may select one of the options, or he can decide not to use any of them. For example, an inherently unstable system can never reach steady state, and hence none of the options would work.

In the automatic mode NESS reduces the rates of integration, increases the final time by 10 percent, and reruns the simulation. An example of a rule for this strategy is:

IF [The user wants automatic mode]

and [the simulation has yet to reach a steady-state]

THEN [Teduce the integration rates by 4.0]

and [increase the simulation final time by 10%].

4.3.3. Stable Simulation

A stable simulation is one which has reached steady state. In the manual mode NESS informs the user of this decision and asks him whether he agrees with this result. If the user feels that the simulation needs a longer period of time, NESS will change the status of the simulation to Not-Yet-Reached-Steady-State and will invoke the strategy_rb rule base to deal with this.

In the automatic mode, NESS tries to ensure the accuracy of the simulation results by reducing the rates of integration by a factor of 4 and re-running the simulation. The results of the two runs are compared. If the difference between the outputs is within specified bounds, the simulation is judged to be accurate. Otherwise, the integration rates are once again reduced by 4, and the simulation is rerun. This process continues until the simulation is judged to be accurate or the number of reruns has gone up to 5. After 3 reruns without achieving the accuracy criterion, NESS asks the user to approve further reductions in the integration rates. Once accuracy is achieved, NESS checks for user-entered performance specifications, If present, NESS compares them with the simulation results, and if the specifications are not satisfied, a compensator is designed and recommended. If the user has not entered any performance specifications, the simulation results are checked with default specifications, and a compensator is recommended if the performance is not satisfactory relative to the default values. An example of a rule which implements this strategy follows:

IF   [The user wants automatic mode]

and   [the simulation has not yet been proven to be accurate]

THEN [Reduce the integration rate by 4]

and   [rerun the simulation]

and   [compare the results of the last two rates]

5. COMPENSATOR DESIGN FACILITY

NESS will suggest a compensator for a stable simulation if the user wishes to improve the system behavior. Two types of compensators, phase lead and phase lag, are available in NESS (14). The following rules will be applied to choose the proper type of compensator.

1) If the system shows a good steady-state accuracy but an unsatisfactory transient response, a phase lead compensator will be designed and recommended to the user.

2) In the case of the system exhibiting satisfactory transient response characteristics but unsatisfactory steady-state characteristics, a phase lag compensator will be designed and recommended.

NESS will use the desired system performance to compute the compensator pole, zero and cascade amplifier gain. If the desired system performance is expressed in terms of time-domain specifications, the compensator will be designed using the root locus technique. If the specifications are in the frequency domain, Bode plot techniques will be used for the design.

6. SYSTEM EVALUATIONS

The evaluation of NESS has been based on: (1) the simulations of three test models, (2) the compensator designs for the test models, and (3) the ability of NESS to interact with the user (4). The coupled system must be able to perform each to these functions in a satisfactory manner in order to be considered acceptable.

In the simulation of the test models, NESS has demonstrated that it can successfully link an expert system with a Fortran simulation package and execute the simulation. The numerical subsystem of NESS performs the numerical integration and uses no knowledge about the system being simulated except what it can infer from the simulation results. NESS is able to determine the proper value of integration time step to produce a stable and accurate simulation of the unknown system without user intervention. NESS is also able to determine performance characteristics of the system, such as percent overshoot or phase margin, from the simulation results. These system characteristics can be compared with user-specified or default values, and a decision made as to whether or not the performance is acceptable. The capabilities of NESS to perform these tasks is shown by the fact that the results obtained from the three test systems compare favorably with analytic results.

The results obtained when the compensator design techniques were applied to the first two test models were very encouraging. Phase lead and phase lag compensators were designed to produce simulation results for the compensated system which were close to analytic results in most cases.

The only major difference between the expected and simulated results was for a phase lead compensator designed for a very lightly damped system. The pseudo open-loop frequency response routine gave results which had a large, but conservative, error relative to the desired results.

Much work needs to be done to make the compensator design procedures more flexible and intelligent. The task of designing compensators when there is no knowledge of the system structure is expremely difficult. If input/output structural knowledge was given, system indentification algorithms could be executed during the simulation in order to determine the parameters. This would graetly improve the capabilities of NESS in compensator design, allowing different types of compensators to be designed.

Although NESS does not have a natural language interface, the menu input format allows a novice user to execute and analyse a simulation with very little effort. Since the user has only a few choises from which to select at each step, there is less chance for confusion and error in data entry.

7. CONCLUSIONS

This chapter has described and illustrated the capabilities of the prototype coupled expert system NESS. One of the major goals of this project was to demonstrate the feasibility of applying expert systems technology to the area of digital simulation of dynamic systems. It has been demonstrated that an expert system can interact with the user to determine the simulation goals, instantiate parameters with the proper values, execute the simulation, and interpret the results relative to performance specifications. It has also been shown that it is feasible for an expert system to intelligently select integration methods and time steps for the simulation, even when little is known about the system to be simulated. This is a difficult task which requires experience and an iterative approach even for a human. Additional work needs to be done in this area to improve the robustness of the knowledge.

NESS has also shown itself to be capable of designing simple compensators to improve the performance of the system. The ability to design appropriate compensators would be greatly improved with more information available about the system being simulated. Additional knowledge needs to be added to expand the capabilities of the compensator design procedures and to allow NESS to handle a wide variety of simulation programs. Future work is being considered to allow NESS to operate in a network environment controlling several simulations.

8. REFERENCES

1. D.A.Waterman, A Guide to Expert Systems, Addison Wesley, 1985.

2. IEEE Computer Society, Proceedings of the Second Conference on Artificial Intelligence Applications, "The Engineering of Knowledge-Based Systems", Minami Beach, Florida, December, 1985.

3. IEEE Computer Society, Special in Expert Systems in Engineering, COMPUTER, Volume 19, No.7, July 1986.

4. K.Kawamura, G.O.Beale, et al, Research on an Expert System for Database Operation of Simulation/Emulation Math Models, Phase II Results, NASA Contract #NAS8-36285, Center for Intelligent Systems, August 1986.

5. J.S.Kowalik, (Ed.) Coupling Symbolic and Numerical Computing in Expert Systems, North-Holland, Amsterdam, 1986.

6. C.T.Kitzmiller and J.C.Kowalik, "Symbolic and Numeric Computing in Knowledge-Based Systems", in Kowalik, J.S. (ed.), Coupling Symbolic and Numerical Computing in Expert Systems, Elsevier Science Publishers, Amsterdam, 1986, pp.3-4.

7. C.T.Kitzmiller and J.S.Kowalik, op. cit., pp. 7-9.

8. G.Vachtsevanos and K.Davey, "Fault Diagnostics for the Space Station Thermal Control System Using a Hybrid Analytical/Intelligent Approach", Proceedings of the IEEE International Symposium on Intelligent Control 1987, pp.54-58, January 1987.

9. A.Valla, "Hybrid Knowledge-Based/Analytical Control of Uncertain Systems", Proceedings of the IEEE International Symposium on Intelligent Control 1987, pp.59-70, January 1987.

10. J.K.Foderaro, K.L.Sklower and K.Layer, The FRANZ LISP Manual, Regents of the University of California, June, 1983.

11. H.S.H.Sabdell, "GENIE User's Guide and Reference Manual", Department of Electrical and Biomedical Engineering, Vanderbilt University, Technical Report #84-003, Nashville, TN, July, 1984.

12. G.O.Beale and T.T.Hartley, "Stability Considerations: Numerical Methods and Control Theory Equivalences", IEEE Transactions on Industrial Electronics, Vol.34, No.2, pp. 180-187, May 1987.

13. J.D.Lambert, Computational Methods in Ordinary Differential Equations, John Wiley and Sons, Chichester, 1973.

14. B.C.Kuo, Automatic Control Systems, 4th Edition, Prentice-Hall, Englewood Cliffs, pp.340-352, 1981.

9

CASE STUDIES IN KNOWLEDGE BASED FAULT DIAGNOSIS AND CONTROL

Dave Reynolds and Chris Cartwright

Cambridge Consultants Limited, England

1. Introduction

The knowledge based approach to design of machinery control and monitoring systems holds out great promise for the future. We can expect to see controllers which can handle difficult processes which have proved impossible to model well enough for current control techniques, condition monitoring systems which integrate a much richer set of data sources than current systems; fault diagnostic systems which can handle complex evolving fault situations without overloading the operator with irrelevant data.

Many of the techniques which will make such applications possible are still in the research stage. Much work needs to be done in refining these techniques and evaluating and classifying where they are appicable so that the process of designing knowledge based controllers can be placed on a firm engineering footing.

Despite the research still to be done, many applications can be both developed and delivered using today's technology. Since the early 1980's a team at Cambridge Consultants limited have been applying state of the art AI technology to practical engineering problems in areas such as machine control, condition monitoring and fault diagnosis. In this paper we describe some of our experiences from this work and discuss the general principles which underly these applications. We support this more general discussion by descriptions of three practical applications we have developed which illustrate a good range of such engineering applications.

2. KNOWLEDGE BASED MONITORING AND CONTROL

Before discussing our approach and experiences it is worth reviewing what we mean by knowledge based control and why we consider it to be an important approach.

Our general aim is to apply the technology of knowledge based programming to problems of industrial control, monitoring and diagnosis. The *knowledge* that we wish to represent is in two rather distinct forms. One the one hand there is empirical knowledge possessed by human experts.

On the other hand there is rigorous knowledge based on the physical structure of the machinery in question and the underlying principles involved. Expert systems technology has been evolved in an attempt to codify and use this first, empirical, class of knowledge. However, in engineering applications the physical structure of the machinery and its design and operating manuals are equally important sources of information. The essence of the knowledge based approach remains that of keeping this information (whatever its source) as explicit as possible so that the resulting programs are easy to understand and easy to maintain.

Why do we believe this to be a valuable approach to engineering problems? What benefits might accure from pursuing it? We can indentify a number of separate benefits which are already apparent in our current applications. Of course, a single given application will not necessarily exhibit all of these benefits.

(1) New applications

Firstly the knowledge based approach allows us to tackle problems which have proved difficult for the more conventional techniques—either through the complexity of the task or though the task being too ill understood.

This is particularly apparent in cases where the underlying process has proved impossible to model mathematically - not uncommon in many process control applications ranging from cement manufacture to pharmaceuticals. Often in such applications the process is being controlled manually by a group of experienced human operators. There have been a few notable successes in codifying this empirical knowledge in the form of a rule based controller. Perhaps the best known example being the Blue Circle cement Kiln controller (1).

There are also applications where the process being monitored is well understood but where the application is too complex to be feasible or economic using conventional approaches. A good example of this is perhaps process alarm handling. In many cases a rigorous causal model of a plant can be constructed which identifies fault cascades and which could support an alarm handling system which separates out primary faults from consequent secondary indications. The difficulty here is not the need for empirical human knowledge but the sheer complexity of the task given the huge number of alarms in a modern process plant. The knowledge based approach can help to manage this complexity by providing tools for representing the underlying causal model on which the alarm analysis can be based. Of some importance for future systems will also be work on distributed intelligent agents to support distrubuted alarm handling in a well structured manner.

(2) Increased flexibility and robustness

Secondly, the knowledge based approach can lead to more flexible and robust systems. In making the underlying knowledge more explicit knowledge based techniques make it possible to re-use that knowledge in different circumstances. In particular, by using explicit representation of control plans we are able to adjust those plans in the face of error conditions to minimise the impact of a fault on the overall objectives of the plant.

They also provides some techniques for taking into account qualitative information from the wider context in which the control/monitoring task is running and so improve robustness in the face of external events. We shall see an example of this later where we used a knowledge based supervisor to control the actions of a range of underlying algorithmic controllers to increase the overall operating range of a piece of machinery.

(3) Improved performance

Even in cases where a purely algorithmic approach is feasible it may still be possible for a knowledge based approach to yield improved performance (such as reduced energy usage). The notion here is that a knowledge based controller may be able to use a wider range of data sources, including qualitative and uncertain data in making its control decisions by exploiting AI based techniques for data fusion and reasoning under uncertainty. We see an example of this in the first case study.

(4) Cognitive matching

Some of the applications in this field are destined for purely embedded operation and need only communicate with data soutces and control interfaces. Others need to interact closely with a human operator. In these areas where the control/monitoring system needs to work in partnership with human operators it is vital that the system provides the operator with a clear picture of what is happening and does not simply deluge him with data. In the future we would see work on user modelling being particularly important in allowing us to construct monitoring/ control systems which provide the operator with information at precisely the correct level of detail for his actual task and in a form and context which enables him to assimilate it rapidly. Even in the short term the benefits of a knowledge based controller in terms of one's ability to understand its line or reasoning and its ability to present data in context can be very beneficial.

We shall see later an example of this in a fault diagnosis system which must interact with an operator in a very controlled fashion. If the system and the operator disagree on a diagnosis the system must be able to rapidly present a very succinct explanation of the differences between the two points of view in order to convince the operator to carry out the recommended control actions.

(5) Reduced cost of ownership

Finally, even if none of the above benefits were relevant and all applications could be handled completely satisfactory by current algorithmic methods there would still be the question of economics. A well designed knowledge based program should be easier to maintain and easier to re-use in new circumstances than a conventional program. Given the soaring costs of software production it may be that the economics of software ownership will eventually be the strongest force behind the move to knowledge based approaches.

At present this list is optimistic. The signs are there that these benefits can often be obtained. However, we cannot yet take a problem definition and quantify the extent to which these individual benefits can be obtained for a particular application without actually trying it out. Much work remains to be done at a research level on many aspects of this technology to both increase its range of applicability and to quantify that range. Knowledge based control is in its infancy. The message of this paper is that despite these limitations we can successfully develop and deliver useful applicatios now.

3. OUR APPROACH

We have already pointed out that there does not currently exist a structured design methodology for knowledge based control/monitoring (KBC) systems. However, from the applications which we have developed to date we can extract some common themes.

One common theme is the general applicability of the control structure depicted in figure 1.

3.1. Monitoring and execution

The essential concept belind this architecture is that the KBC maintains an explicit detailed control plan and a model of the current state of the system. For most of the time the diagnosis and planning modules are quiescent and the monitor/execution module is following the control plan. It monitors the changing state of the plant being controlled both to detect unexpected external events and to ensure that the control actions are successful.

Some faults will be detected by fault-specific monitors. For example, a power imbalance between two linked drive shafts can be detected as a fault, independent of the control situation. Other faults are detected by their impact on the control plan. For example, a stuck fluid control valve may only be observable as an inconsistent set of fluid flow readings during particular control sequences.

The essential thing to note about KBC applications is that this monitoring module is monitoring an evolving *process* not a static system. The representations for the control plan, the control effects and the system state must thus capture this evolution very clearly. In the future we might anticipate that research in representations such as temporal logics[2,3] and qualitative physics[4] might be applied. At present we have found the most useful representation to be that of a *script*. A script, in this context, is simply a time ordered sequence of actions with explicit representations for the events which trigger the next action and the assumptions which underly the action (for the purposes of the planner). This has some ties with the notion of a script used in discourse understanding[5] and with work on recognition of inherently temporal processes[6]. The precise nature of the representation of event triggers,

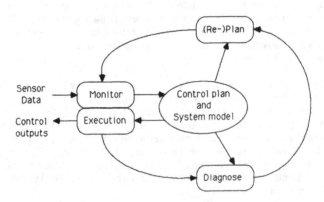

Fig. 1 An architecture of KB control.

actions and assumptions depends on the application.

Once a problem has been detected then the current control plan will normally be suspended and the diagnosis module will be invoked to determine the nature of the problem.

3.2. Diagnosis

The actual diagnosis process depends upon the particular application. The diagnosis module has very clear information on the context in which the fault occured which simplifies the diagnosis process. For example, a fault which is detected by inconsistent results following a control action is immediately localised to the components (or sensors) affected by that control action.

It is extremely important to take into account the continuous reasoning nature of KBC applications. Typically the sensor data will be continually changing as legitimate control actions are carried out and it is important to distinguish between true faults and events which are simply due to transient response to control actions. The faults themeselves will develop over time, both in terms of severity and in terms of triggering other, secondary, faults. The importance of an explicit system state model with history is obvious here.

Furthermore, we generally only have partial observability of the system being monitored. Thus when a fault indication occurs we will rarely have sufficient data to arrive a full diagnosis and we need to in some way perturb the system being monitored to distinguish between alternative faults. This gives rise to two problems. Firstly, that perturbation must be carefully chosen so as not to adversely affect the current operation of the plant more than is necessary, often the KBC must convince a human operator that a particular diagnostic check must be carried out. Secondly, we have to be able to reason with partial information. It is important to be able to determine the class of possible faults and the list of most likely candidate faults in the absence of full information in order to decide how important the malfunction is and what scale of perturbation is justified in further refining the diagnosis.

One of the central pragmatic lessons which we wish to relate in this paper is the importance of chosing the knowledge representation so that it reflects some of the syntactic structure of the underlying problem. The more common notions of representing diagnostic knowledge as heuristics and thus using rule based representations has generally proved ineffective in our applications to date. It has proved difficult to capture with rule based methods the temporal notions involved in faults which develop over time and which are hidden by other legitimate changes in the behaviour of the target machinery. Even a notion as simple as that of scripts can greatly improve the intelligability and robustness of a KBC application. The current work on diagnosis from first princimples (for example see[7]), whilst very important for static diagnosis does not yet handle these aspects of continuous reasoning fully satisfactorily.

3.3. Planning

When a fault has been isolated the planning module is invoked to find a way round it, if possible. It is usually impractical to use a general purpose planning system here due to the severe time constraints on the planning - we must decide what to do about any fault fast, before any physical damage is done. For some of our KBC applications it is possible to keep the original control plan available in a form which is easy to

modify in the context of a given fault. In effect we precompile our
domain knowledge (manually) into a set of partial plans with explict
assumptions. The replanning task just needs to check which plan components
are invalidated and look for alternative partial plans which would fit the
changed circumstances.

4. CASE STUDIES

To explain the way in which this common architecture theme evolved
and to illustrate how it can be applied in practice let us look at a range
of control and monitoring applications which we have constructed over the
last few years.

4.1. A supervisory control application

Our first case study serves both to illustrate this overall
architecture in a machine control context and to show one way in which we
can mix conventional and knowledge based control approaches to their
mutual benefit.

4.1.1. The application

The application is that of dynamically positioning a newly designed,
highly manoeuverable ship despite wind and tidal forces. The ship in
question is equipped with fully steerable Voith-Schneider stern propulsors
and bow thruster tubes. The control engineering group at Cambridge
Consultants had already assisted in the development to two different
control algorithms for this ship. One algorithm was a dynamic positioning
(DP) algorithm based upon multi-variable control theory. The second,
position control by manoevering (PCM), worked by balancing the ships's
stern thrusters against the environmental forces in a manner akin to that
of balancing an inverted pendulum.

Each of these algorithms worked very well in part of the operating
range of the ship but no single algorithm (and no single setting of the
primary control parameters for the algorithm) was entirely satisfactory
over the complete desired operating range of the vessel. The aim of the
AI project was to investigate the potential of a knowledge based supervisor to control, in real time, the selection and parameter setting for
these algorithmic controllers. Thus leading to the control structure
shown in figure 2.

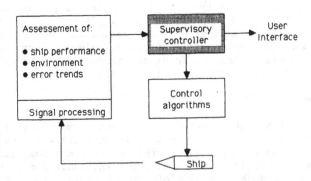

Fig. 2 Structure of the ship positioning application.

This is a real-time, embedded knowledge based application; "embedded" both in the sense of being largely non-interactive, and in the sense of being closely integrated with other software systems. We were thus working under the severe constraints of needing to be able to react to events automatically at very high speed with relatively constrained hardware resources. The work to date has all been carried out using a detailed, real-time hydrodynamic simulation of vessel provided by our clients.

In this case the source of the expertise or knowledge which was to form the basis for the supervisor was the control engineers themeselves who designed the algorithmic controllers and understood their capabilities and limitations - not the pilots who were faced with carrying out the control task manually.

4.1.2. The solution - Knowledge Representation

As we pointed out above the key element in devloping such an application is the selection of the representations to be used for the system model and the control knowledge. After an initial, unsuccesful trial with a rule based formalism for representing the control knowledge we chose a variant on the script based approach metioned above.

The overall scheme for controlling the vessel was divided into seven different major regimes. For each control regime we developed a script containing:

(1) the control plan itself (i.e. the control algorithm and parameter settings to use),

(2) the assumptions on which the control plan were based,

(3) rules for how to adjust the control plan if one or more assumptions were violated.

Partitioning the problem into these destinct control regimes greatly simplified the task of ensuring that all control states were suitably covered. It also gave tremendous benefits in terms of efficient implementation.

4.1.3. The solution - Architecture

The architecture which implements the supervisory decision module closely matches the general scheme outlined above and in shown in figure 3.

At any time the controller's understanding of the ship status is represented in semi-quantitative terms by a state vector generated by the signal processing and interpretation modules. The monitor module/execution module simply selects the control algorithm and parameters determined by the current control script and then monitors the changing information in the state vector against the assumptions behind the current plan. The current plan script defines very clearly what information needs to be monitored leading to an efficient implementation of this crucial part of the architecture.

There is essentially no diagnosis to perform. The underlying fault we need to determine is which assumption underlying the control algorithm has been violated by the changing conditions. The monitor itself detects these violations directly because the assumptions are made explicit.

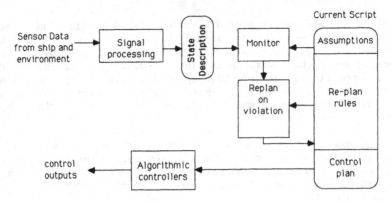

Fig. 3 Structure of the Supervisory Controller.

The replanning task consists of switching to a new control script. This uses the assumption violation rules of the current control script to determine which new state (control regime) to switch to. Thus it is the responsibility of the current script to determine which new one should take over. There is no question of a scripr matcher failing and leaving the ship without a controller – although one script alternative is to hand over control to the human pilot. The replanning rules includes a simple notion of hysteresis in that the switch from say script 1 to script 2 will occur at a different point from the switch from script 2 back to script 1.

Again the use of explicit assumptions here suggests a very efficient implementation. The replanning rules are indexed by the assumptions which are being monitored. A single script may have around ten major assumptions associated with it but for a given assumption only a small subset (generally less than six) rules is relevant. Thus the replanning process directly indexes into a small subset of the total ruleset and can very rapidly determine the best new control script to employ.

4.1.4. Discussion

This example application illustrates many of the points we are trying to make. In AI terms the technology employed is simple but powerful. The script based representation is natural for the particular knowledge in question. The overall control cycle we discussed before works well in constraining the work required at run-time. Very little search is needed because each operation is carried out in the context of a very specific set of assumptions or violations. This leads to a natural and very efficient implementation which reflects the underlying structure of the problem.

One point to note with this application is that the decision module is not simply another control module which could have been designed more conventionally because it is using information on the environment and ship status which is inherently qualitative and uncertain in nature. We have no formal model for how this information can be exploited but are relying on the experise and intuition of the control engineers who developed the original DP and PCM algorithms.

For more details on this work see[8].

4.2. A Fault Monitoring Application

This project was very much concerned with the monitoring and diagnosis phases in the overall model shown above, and was designed as a decision support system rather than an autonomous controller. Neverthe-less some common themes can be seen with the more control - oriented systems, particularly in relation to the need for a well-structured knowl-edge representation tailored to the domain at hand.

The particular application was selected on pragmatic criteria ralating to its performance as a demonstration system: a well constrained and relatively well understood problem domain, available expertise, test-ability and some important user MMI issues to be resolved.

4.2.1. The Application

The chosen application was online fault diagnosis on board an air-craft. The problem is well understood and testable due to the existance of detailed diagnosis information available to the pilots in written form. This assists the pilot in his diagnosis task, based on the observ-able symptoms: warning lights, unusual analogue instrumentation readings and performance degradation.

Key areas of interest centre around the issues of continous reason-ing:

Fault recovery and multiple fault handling are necessary in any system which is to function in an online environment. Also some notion of assumption-based reasoning is likely to be required for tackling issues of failed sensors or instrumentation. This would allow malfunctions to be hypothesised without committing to the consequent inferences.

Understanding normal control inputs, and distinguishing the consequences from failure modes is another important area. The aim here is equivalent to a car which, when it stalls, doesn't immediately tell you that your oil pressure has gone critical, your brakes have failed, and so on, but expects these symptoms as a consequence of the stall. Subsequent diagnosis of faults with some subsystems in non-standard mode (i.e. isolated, or turned off) requires a good model of which symptoms are affected by such actions, and thus what the expected set of symptoms is for a given fault in the new state.

Interaction with the human agent (in this case the pilot) is often necessary to perform diagnosis in a decision support system such as this. An empirical phase of "try this and see what happens" is frequently required for unambiguous diagnosis. In a fully autonomous controller, the required perturbations to the aircraft state could be administered direct-ly by the system; without this power, the IKBS must liaise with the pilot in a flexible and responsive manner to gather the required extra data.

4.2.2. The Solution - Rule Based Approach

Two approaches to this problem were tried, the first being a fairly straight forward rule based system. The implementation language, the MUSE Toolkit[9], allows a production ruleset to run in a continuous monitoring mode, thus providing a simple event-driven diagnosis system. The action of the ruleset was to modify a number of fault hypothesis structures: effectively partitioning the set of possible faults for a given subsystem as new evidence arrived. Elimination of the candidate fault set down to a small number of options provoked an interactive diagnosis phase with the pilot on the basis of discriminators between possible faults, If and when the pilot chose to perturb the system in an appropriate manner, diagnosis could be completed and consequent advice or recommendations given.

Considerable effort was also devoted to an explanation system based around the hypothesis structure and discriminant symptoms. Thus the pilot was able, with minimal interaction, to request explanations either of why a given fault had been diagnosed, or why it had been chosen in preference to the pilot's preferred candidate. These explanations were given at the level of symptoms and discriminators, *not* rule firings.

4.2.3. The Solution – Model Based Approach

The second approach was the result of a critical review of the implementation above, which has serious fundamental limitations in respect to several of the points outlined above. Many of these limitations can be traced to the use of a rule-based approach. Despite its normally being thought of as a declarative representation, the limitation of a rule-based approach here is due to the "compiled" nature of the knowledge in the rules. Owing to the heavy emphasis on syntactic model knowledge in this type of system (rather than heuristics), it is perfectly possible to have a representation of fault-symptom relationship which is transparent at run time, allowing much more flexibility in reasoning about the inter-relationship themeselves. Thus each fault, symptom and fault group can be represented in a graph structure defining the relationship of required and optional symptom sets for given faults. Considerations such as sensor failures and response to normal operation make such a transparent archi-tecture mandatory.

This second architecture was based around a syntactic (surface) model of the fault-symptom relationship. The resulting graph acts boths as a static representation of the relationship and as a dynamic model of current state. The reversibility of graph state changes guarantees that any given combination of symptoms will always give rise to an identical graph state, however reached. This gives the vital ability to cope with an on-going fault situation, including recovery and intermittent problems. A further aspect of the system is needed to cope with transient effects (e.g. fault occurs – pilot alters controls – fault disappears), which are lost in such a dynamic graph structure. The approach here was to use a "history" list of significant events emerging from the graph state. This then gives a sequence of occurrances on a temporal axis which can be matched for patterns indicative of known fault development paths.

4.2.4. Discussion

Probably the most important point to emerge from this project is the importance of maintaining sufficient transparency of knowledge as is appropriate to the type of knowledge being used. Whereas heuristic knowledge by its nature cannot be "structured" any further, knowledge characterising well-defined physical or causal links should be represented so as to allow access to those interralationships by the running system. This observation offers support for the growing interest in model-based AI, even for syntactic level representations of physical systems. Rules should be used with care and where appropriate.

A more general point concerns the care with which such a system must be designed if it is to perform usefuly in a continuous environment. The need for a specific real-time approach to diagnosis has to appreciated from the beginning if a usable system is to emerge. This requirement is of course equally true for a diagnosis module forming part of an integrated monitoring and control system.

4.3. A Knowledge based Control Application

The final application we consider encompasses the entire spread of the monitor-diagnosis-control model discussed above. Since many of the ideas have already been introduced, we restrict ourselves here to a brief description of the domain, and some comments on how this apllication relates to the points made in section 3.

4.3.1. The Application

The Application chosen for this system concerns flow control in a complex fuel supply system (also coincidentally in an aircraft). The aircraft concerned has around a dozen fuel tanks, distributed around the fuselage. The aim of the system is to drain these tanks so as to maintain a constant supply to the engines (needless to say), whilst also attempting to satisfy a number of other constraints:

Preservation of the center of mass of the aircraft: a number of tanks are significantly distant from the central axes of the aircraft. These must be drained at the correct times, and in parallel, to preserve the balance of the craft within set limits.

Draining tanks in or close to a defined order: some tanks can be jettisoned when empty to decrease weight, others improve stability when empty. A specified order of draining is laid down to be adhered to in the absence of overriding contraindications.

Minimising any losses due to system faults: an intelligent modification of the drain order or paths (of which there are several for each tank) can save fuel after a fault. For example, a leaking tank should be drained at the earliest opportunity to minimise fuel loss from the leak.

4.3.2. The Solution

The solution implemented follows the scheme laid out in section 3 quite closely.

The system begins with a default plan embodying the preferred drain order. This plan is converted into a sequence of valve operations, and executed in a simple cycle of three stages; open valves, wait for termination condition(normally a tank drained to a given level), and the close valves (or some combination of open and close requests to move to the next drain path in the sequence). If nothing untoward happens, this sequence is pursuel until the tanks are empty.

If a fault condition is spotted, the diagnosis module is invoked. Faults are typically valves or pumps which dont't open (start) or close (stop) when requested. Each time requests are issued the system check for compliance. A problem here could either be due to a component failure or a sensor failure on that component. The other class of fault, a leak, is spotted by unexpected drain rates, monitored by time-conscious system demons. Diagnosis is straightforward in this system and occurs via a simple table lookup system, based on the notion of intersecting sets of fault-to-symptom mappings (compare with[10]).

On isolating the fault, the planner redesigns its drain plan to work round the fault. This will involve either modifying the order of draining, or changing to a new route. The former is used in the case of a leaking tank mentioned above. The latter is useful if a valve is stuck open, allowing a tank to drain which should not. Altering the route can effectively "close" this valve by blocking its drain route downstream. The new plan is checked against the top-level constraints before being put into action.

The system is build around an object-oriented model of the fuel system, which serves not only as a representation of the current believed state of the system, but also an active representation tied to a simulation of the actual fuel system state.

4.3.3. Discussion

The structure of this application reflects closely the model outlined earlier. Note particularly the need for dynamic re-planning of the script structure, and the use of a central system model used by all reasoning agents. This is a direct control system, not a supervisory controller. This is possible in this case due to the simplicity of the control actions required and the relatively slow time constants for the system.

5. CONCLUSIONS

We have described a range of knowledge based control and monitoring applications and have attempted to highlight some of the common themes underlying our approach to them. This work has been generally pragmatic in outlook and so the observations we have made are purely empirical. Never the less we hope that they may be of use to other engineers in the construction of practical knowledge based control applications.

6. ACKNOWLEDGEMENTS

We would like to thank all of our colleagues in the AI and control engineering groups at Cambridge Consultants. Especially Martin Bennett, Rob Zanconato, Bryan Clarke and Charles Boulton; whose work formed the basis for this paper.

We would also like to thank our clients for providing such stimulating problems. The first case study was sponsored by the Royal Aircraft Establishment at West Drayton. The second case study was sponsored by the Human Engineering Department, Flight Systems Division of the Royal Aircraft Establishment at Farnborough. The third case study was a cooperative programme with the British Aerospace Military Aircraft Division at Warton, England.

7. REFERENCES

1. D. W. Haspel, J. C. Taunton, Application of Rule Based Control in the Cement Industry, in: *Proc. Expert Systems and Optimisation in Process Control*, (1985).

2. J. F. Allen, Maintaining Knowledge about Temporal Intervals, *Com. ACM*, V 26, 11, 832-843 (1983).

3. R. Turner, *Logics for Artificial Intelligence*, Ellis Horwood, Chichester (1984).

4. K. D. Forbus, Qualitative Process Theory, MIT AI-TR-789, 1984.

5. A. Barr and E. A. Feigenbaum, *The Handbook of Artificial Intelligence*, Volume1, Pitman Books, London (1981).

6. R. McArthur, J. R. Davis, D. E. Reynolds, Scenario-Driven Automatic Pattern Recognition in Nowcasting, *J. Atmosphere and Oceanic Technology*, (1987).

7. R. Reiter, A Theory of Diagnosis from first Principles, AI V 32, 57-95, (1987).

8. D. E. Reynolds, C. B. Boulton, S. C. Martin, AI Applied to Real Time Control: A Case Study, in *Applications of AI in Engineering Problems*, pp. 573-583, Springer - Verlag, New York (1986).

9. MUSE User Manual, Combridge Consultants Limited, England, 1987.

10. J. A. Reggia, D. S. Nau, P. Y. Wang, Diagnostic expert systems based on a set covering model, *Int. J. Mac-Machine Studies*, 19, 437-460 (1983).

10

CONTROL ALGORITHM ACQUISITION, ANALYSIS AND REDUCTION: A MACHINE

LEARNING APPROACH[+]

Wojciech Ziarko* and Jack David Katzberg**

*Department of Computer Science, University of Regina
 Regina, Saskatchewan, CANADA S4S 0A2

**Electronic Information Systems Engineering
 University of Regina, Regina, Saskatchewan, CANADA S4S 0A2

1. INTRODUCTION

This chapter presents a procedure for capturing process operators supervisory control methods. Present measurements and supervisory control actions are recorded. This information is then analyzed to determine the relationship between measurements and the operator control actions. The results of the analysis are subsequently used to synthesize a control algorithm to substitute for the operator or advise the operator which control action were likely to be best, in a given situation.

1.1. Motivation

Many industrial process, such as cement kilns and crude oil stills, are not completely understood. The complex reactions have never been completely modelled. Other systems such as power systems have models which are too complicated for control system design. These systems have been operated satisfactorily for many years under operator supervision. Many conservative managers are very reluctant to move to a completely automatic control system. In any case the costs of modelling or identifying the systems and designing a supervisory computer control system would be very large.

In this chapter a method a computer can use to learn and mimic the operators supervisory control methods is given. The approach is automatic and the system is transparent to the operator. Once the control methods have been learned they could be used to advice less experienced operators. When management and the operational staff are convinced the advice is reliable, the computer based algorithms, learned from the operators, could control the system directly.

+ This research was supported in part by a grant from Natural Sciences
  and Engineering Research Council of Canada.

1.2. Approach

The mathematical formulation of the problem will be stated precisely and illustrated with examples computed using qualitative data analysis and reduction systems ANLYST developed by the first author. Then the relevant elements of theory of rough sets will be introduced (1,2,3). It will be showm how the rough set approach can be used to capture the control methods used by process operators. Methods will be developed to produce a minimal set of dependencies linking the measurements and past control actions with the present control decisions. This will reduce the computational burden when the learned control procedures are applied as an advisory or direct control system.

As it has been suggested adaptive control systems could be used to replace operator supervisory control, adaptive control methods will be surveyed (4,5,6). Self-optimizing supervisory control systems(7) are those best suited to the problems the machine learning system can handle. A number of application areas will be discussed and the advantages and limitations of adaptive and expert system control methods will be contrasted and compared with the machine learning approaches developed in this chapter.

2. SYSTEM AND OPERATOR MODELS

2.1. Basic Model of the Control System

The control system has two main components,the system being controlled and the system operator. The operator is assumed to be an experienced individual, an expert, capable of selecting correct control actions in response to the information available to him for a particular event or situation. As the operator makes control actions at discrete times as situations or events arise it is natural to assume the system obeys a model of form

$$\mathbf{x}_k = f(\mathbf{x}_k, \mathbf{m}_k) \tag{1}$$

$$\mathbf{c}_k = g(\mathbf{x}_k, \mathbf{m}_k) \tag{2}$$

where $\mathbf{x}_k$ is a state vector of n elements, $[x_1, x_2, \ldots, x_n]^T$
$\mathbf{m}_k$ is an input vector of p elements, $[m_1, m_2, \ldots, m_p]^T$
and $\mathbf{c}_k$ is an output vector of q elements, $[c_1, c_2, \ldots, c_p]^T$.

The elements of the input vector, $\mathbf{m}_k$, are the control variables the operator uses. These may be naturally discrete actions such as switching equipment on or off or actions which might, using a conventional design approach be considered continuous. We will assume these continuous variables are broken into discret ranges natural to the operator. For instance if a flow was being controlled if the flow was good he would leave the valve set as it was, if it was too slow he would open the valve more, and if it was too fast he would close the valve a bit.

Similarly, the elements of the output vector, $\mathbf{c}_k$, are the measured variables the operator uses. These may be naturally discrete such as lights on or off or measurements, such as temperature, which are continous. These continous variables are broken into discrete ranges natural to the operator. For instance, good, if it was quite close to the desired value, a bit high or a bit low, if the product produced was still within specifications, or unacceptably high or unacceptably low, if the product was out of specification. That is, in the rough set approach, we are not interested in precise, continuous readings as is usually the case, but in meaningful, discrete ranges of values.

The state vector and the model form are not available to the operator and will not be used. For simplicity it will be assumed as in (9) that all the information the operator has available to him, in a given situation, are the present measuremtnts. Methods will be derived for automatically producing from past recorded actions the operators decision rules for selecting control actions based on value of the output vector.

The approach that is presented is a type of output feedback. Using a conventional control theoretic approach one would use a given number of past measurements and control actions to compute state if the performance of output feedback controller was not acceptable, so state feedback could be used. The operator has no direct knowledge of the complete state vetor, however he may use past measurements and control actions in selecting present actions. It is a straight forward matter to add this extra information when producing the decision rules. If necessary one can generalize the approach presented there to extract the operators decision rules for selecting present control actions from the present measurements and a given number of past control inputs and outputs.

2.2. Learning Subsystem

Our objective is to analyze the past actions of the system operator versus system output information in order to obtain a generalized control algorithm which could be used to offer advice or effectivaly replace the operator in the future. The control algorithm will have dynamic character i.e. it will be continuously modified to better reflect accumulated experience. In this sense the control system will have the ability to learn in the course of the time as new situations or experiences will be added to the control system.

The approach presented in this paper is anspired by the idea of automatic learning from examples which is being developed in Artificial Intelligence for a number of years[10]. In particular recent results in the theory of rough sets[3] are directly applicable to the problem addressed in this paper. The results of application of rough sets to the problem of expert's knowledge acquisistion[11] can be utilized for control algorithm acquisition. In fact, some experiments along these lines have successfully been performed with cement kiln production[9] algorithm although no background was given in (9) about control algorithm reduction and analysis. It must be emphasized at this point that we are chiefly interested in acquisition of a sort of operator's experience which constitutes his background knowledge. This is the knowledge which accumulated during years of experience in the form of heuristic rules or feel the operator is using to obtain desired system's response. As it is demonstrated by the current practice with expert systems, this kind of knowledge is most difficult to formalize in the form of rules as very often the expert himself is unable to say why a particular decision was made. Therefore, the machine learning approach should be taken to induce decision rules from examples of decision made by an expert, in this case, the operator of the system.

2.3. Representation of Operators Experience

All past and present actions of the system operator along with the system's state and response information wiil be recorded in an information table as illustrated in Table 1.

The Table 1 the representations of thirteen situations $E=\{e_1,e_2,\ldots,e_{13}\}$ requiring operator's decision are included. Each situation e_i is represented by values of measurements $C=\{c_1,c_2,c_3,c_4,c_5,c_6\}$, and the value of an action variable m_1 which reflects operator's response to conditions indicated by condition variables (the measurements will be referred to as condition variables). The content of the Table 1 can be viewed as a representation of operator's experience which will be used as a basis for control algorithm synthesis.

The Table 1, referred to as an Information Table, is a special case of a Knowledge Representation System (or Information System) whose logical properties have been investigated by many researchers[12]. Formally, the Information Table (IT) is a quintuple

$$IT=<E,C,M,V,f> \quad \text{where}$$

E is a set of situations;
C is a set of condition variables, it is the same set of variables that are in the output vector;
M is a set of action variables;
V is a union of domains of all system variables i.e.

$$V= \bigcup_{a\varepsilon C\cup M} V_a \quad (V_a \text{ is a domain of variable a)};$$

and
$f:E \times (C\cup M) \to V$ is an information function which assigns values of system variables to situations. That is, for every situation e_i, $f(e_i,c_j)$ is a particular value from the set or range of possible measuremtnts, c_j, for each j.

Table 1

An Example of an Information Table

SITUATIONS	CONDITIONS					C	ACTIONS M
E	c_1	c_2	c_3	c_4	c_5	c_6	m_1
e_1	1	2	1	2	2	2	1
e_2	2	1	3	1	1	2	2
e_3	2	6	3	1	2	2	1
e_4	1	1	1	1	1	2	1
e_5	1	1	1	1	1	2	2
e_6	3	7	3	2	1	1	2
e_7	2	4	1	1	2	2	1
e_8	2	7	1	1	2	2	1
e_9	2	5	3	1	2	2	1
e_{10}	1	5	3	1	2	2	1
e_{11}	3	1	3	1	2	2	2
e_{12}	2	5	3	1	2	2	2
e_{13}	2	1	2	1	1	2	2

From now on we will use the notation introduced in this subsection to explain the idea of acquisistion and analysis of control algorithm in the context of rough sets.

3. CONTROL ALGORITHM ACQUISITION, ANALYSIS AND REDUCTION: ROUGH SET APPROACH

Before we introduce the notion of control algorithm induced from Information Table some general properties of Information Tables need to be clarified.

3.1. Classification of Operator's Decisions

Based on any subset $B \subseteq C \cup M$ of system variables we can partition the set of situations represented in Information Table into a number of disjoint classes. The classes are constructed according to an equivalence relation $Q(B)$ defined as follows.

We say that two situations e_i and e_j are indistinguishable or equivalent with respect to B if $f(e_i,b)=f(e_j,b)$ for all variables b belonging to B. The set of all events such that this is true, forms an equivalence class, X, a subset of E, the set of all situations.

In the equivalence relation $Q(B)$ defined in this way each of its equivalence classes X is associated a unique combination of values of variables from B. That is each $b \in B$ has a specified value as in Example 1. This combination, referred to as a description of the class X will be denoted by

$$\text{Des}_B(X) = \bigwedge_{b \in B} (b:=f(e,b)) \quad \text{where e is}$$

an arbitrary situation from the class X, the symbol $\bigwedge$ denotes a logical conjunction of terms and := is an assignment symbol. That is, the description is the set of values each variable within B must take on to be associated with a situation in the set X. The collection of all equivalence classes of the relation $Q(B)$ will be denoted by B*.

Example 1

Based on the values of variables $B=\{c_2, c_3\}$ we can produce, from the Information Table, Table 1, the partition $B^*=\{X_1,X_2,X_3,X_4,X_5,X_6,X_7, X_8,X_9\}$ of the set of situations E
with
$X_1=\{e_1\}$ and $\text{Des}_B(X_1)=(c_2:=2) \wedge (c_3:=1)$,
$X_2=\{e_2,e_{11}\}$ and $\text{Des}_B(X_2)=(c_2:=1) \wedge (c_3:=3)$,
$X_3=\{e_3\}$ and $\text{Des}_B(X_3)=(c_2:=6) \wedge (c_3:=3)$,
$X_4=\{e_4,e_5\}$ and $\text{Des}_B(X_4)=(c_2:=1) \wedge (c_3:=1)$,
$X_5=\{e_6\}$ and $\text{Des}_B(X_5)=(c_2:=7) \wedge (c_3:=3)$,
$X_6=\{e_7\}$ and $\text{Des}_B(X_6)=(c_2:=4) \wedge (c_3:=1)$,
$X_7=\{e_8\}$ and $\text{Des}_B(X_7)=(c_2:=7) \wedge (c_3:=1)$,
$X_8=\{e_9,e_{10},e_{12}\}$ and $\text{Des}_B(X_8)=(c_2:=5) \wedge (c_3:=3)$,
$X_9=\{e_{13}\}$ and $\text{Des}_B(X_9)=(c_2:=1) \wedge (c_3:=2)$.

3.2. Control Algorithm Induced From Data

The notion of the control algorithm derived from Information Table was originally introduced in[13] as a formula of more general decision language. Here, we adopt the original idea without introducing the formalism of the decision language.

In principle, the control algorithm induced from the table reflects presence of dependencies existing between two groups of system variables: condition variables and action or control variables. Therefore, all notions and results presented in this chapter are, in fact, applicable to analysis of dependencies in Information Tables.

Let $C^* = \{X_1, X_2, \ldots, X_I\}$ denote the classification of the set of situations E generated by values of condition variables C. Similarly, let $M^* = \{Y_1, Y_2, \ldots, Y_J\}$ be a classification of situations according to values of control variables M. Based on the degree of dependency existing between condition and decision attributes we can distinguish two main classes of control algorithms: deterministic control algorithms where all actions are uniquely defined by conditions and nondeterministic algorithms where for some combinations of conditions more than one action is possible.

Formally, the deterministic control algorithm can be defined as a sequence of deterministic decision rules

$$r : r_1, r_2, \ldots, r_n \text{ where}$$

each decision rule is a conditional statement associated with each class of the partition C^* in the following way:

$$r_i : \text{Des}_C(X_i) \to \text{Des}_M(Y_j)$$

if and only if $X_i \subseteq Y_j$. That is given a vector of output values always implies a given vector of control input values. In other words for all situations where the output is a particular vector value the control action will have a particular discrete vector value. The control action may occur for other situations as well.

The deterministic control algorithm reflects the presence of functional dependency between condition and control variables in Information Table. In general, each decision rule r_i in the control algorithm r should be interpreted as a directive stating that if the conditions specified in $\text{Des}_C(X_i)$ are satisfied by the system being controlled then the actions encoded in $\text{Des}_M(Y_j)$ should be taken.

Example 2

An example of a decision rule derived from the Table 1 is the following statement:

$$(c_1 := 2) \wedge (c_2 := 1) \wedge (c_3 := 3) \wedge (c_4 := 1) \wedge (c_5 := 1) \wedge (c_6 := 2) \to m_1 := 2;$$

or in a less formal way:

IF (value of c_1 is 2) and (value of c_2 is 1) ... THEN take the control action specified by value 2 of the control variable m_1.

To define formally the notion of non deterministic control algorithm the nondeterministic decision rule must be defined first. Let us associate with any class $X_i \in C^*$ a collection D_i of those classes from M^* which have non-empty intersections with X_i, that is,

$$D_i = \{Y_j \in M^* : Y_j \wedge X_i \neq \emptyset\}.$$

The nondeterministic decision rule r_i associated with the class X_i of D^* can be defined as a statement

$$r_i : Des_C(X_i) \rightarrow \bigvee_{Y_j \in D_i} Des_M(Y_j)$$

where the conclusion part of the rule, the right side is a disjunction or union of descriptions of all classes of M^* which are contained in D_i.

The nondeterministic decision rule reflects the fact that there is more than one control action possible in response to conditions specified by $Des_C(X_i)$.

Example 3

From Table 1 the following nondeterministic decision rule can be formed

$$(c_1 := 1) \wedge (c_2 := 2) \wedge (c_3 := 1) \wedge (c_4 := 1) \wedge (c_5 := 1) \wedge (c_6 := 2) \rightarrow (m_1 := 1) \vee (m_1 := 2).$$

The control algorithm r: r_1, r_2,...,r_n is said to be nondeterministic if not all of its decision rules are deterministic. The nondeterministic control algorithm reflects the presence of partial functional dependency between condition and control variables if some of the rules are deterministc, total lack of functional dependency if all of the rules are nondeterministic. However, in the later case there may be still some probabilistic dependency (3) which is not captured by the deterministic rough set model discussed here.

The presence of nondeterministic decision rules indicates that in some situations the control must be turned back to the system operator as it is impossible to decide automatically which action should be taken based solely on measured system parameters. The control program implemented according to such an algorithm will operate a part of the process relying on human operator in situations corresponding to nondeterministic decision rules of the control algorithm.

From the original Table 1 the following control algorithm can be derived:

Example 4

IF $(c_1 := 1) \wedge (c_2 := 2) \wedge (c_3 := 1) \wedge (c_4 := 2) \wedge (c_5 := 2) \wedge (c_6 := 2)$
THEN take action 1;

IF $(c_1 := 2) \wedge (c_2 := 1) \wedge (c_3 := 3) \wedge (c_4 := 1) \wedge (c_5 := 1) \wedge (c_6 := 2)$
THEN take action 2;

IF $(c_1 := 2) \wedge (c_2 := 3) \wedge (c_3 := 3) \wedge (c_4 := 1) \wedge (c_5 := 2) \wedge (c_6 := 2)$
THEN take action 1;

IF $(c_1 := 1) \wedge (c_2 := 1) \wedge (c_3 := 1) \wedge (c_4 := 1) \wedge (c_5 := 1) \wedge (c_6 := 2)$
THEN turn control back to operator;

IF $(c_1 := 3) \wedge (c_2 := 7) \wedge (c_3 := 3) \wedge (c_4 := 2) \wedge (c_5 := 1) \wedge (c_6 := 1)$
THEN take action 2;

IF $(c_1:=2) \wedge (c_2:=4) \wedge (c_3:=1) \wedge (c_4:=1) \wedge (c_5:=2) \wedge (c_6:=2)$
THEN take action 1;

IF $(c_1:=2) \wedge (c_2:=7) \wedge (c_3:=1) \wedge (c_4:=1) \wedge (c_5:=2) \wedge (c_6:=2)$
THEN take action 1;

IF $(c_1:=2) \wedge (c_2:=5) \wedge (c_3:=3) \wedge (c_4:=1) \wedge (c_5:=2) \wedge (c_6:=1)$
THEN turn control back to operator;

IF $(c_1:=1) \wedge (c_2:=5) \wedge (c_3:=3) \wedge (c_4:=1) \wedge (c_5:=2) \wedge (c_6:=2)$
THEN take action 1;

IF $(c_1:=3) \wedge (c_2:=1) \wedge (c_3:=3) \wedge (c_4:=1) \wedge (c_5:=2) \wedge (c_6:=2)$
THEN take action 2;

IF $(c_1:=2) \wedge (c_2:=1) \wedge (c_3:=2) \wedge (c_4:=1) \wedge (c_5:=1) \wedge (c_6:=2)$
THEN take action 2.

3.3. Analysis and Reduction of Control Algorithm

From the point of view of the operator of a complex system it is
always desirable to reduce the amount of information required to take
correct control action. By eliminating the need for taking some state
measurements we can simplify the system and reduce the complexity of the
decision making process. One of the main objectives of the analysis of
the experimental data used to derive control algorithms is to produce
simplified control algorithm without sacrificing the accuracy of control
decisions.
In this analysis we can distinguish two main stages:

— uniform reduction of condition variables which is concerned with elimination of some condition variables from the information table;

— nonuniform reduction of condition variables in which apart from eliminating some condition variables also some values in the information
table are replaced by "don't care" symbols ($\#$).

Similar reduction of the group of control variables can be done.
Because, in reality, none of the control actions can be eliminated for
obvious reasons, the reduction of control variables in the table means
that some control actions can be combined together (for instance two separate switches can be joined together by a horizontal bar) resulting in
a simpler and more efficient system operation, or some control actions
may be left at fixed values.

3.3.1. Variable Dependencies

To describe the method of condition variables reduction we will
adopt the terminology of the theory of rough sets introduced by Pawlak(2).
This method is applicable to both deterministic and nondeterministic control algorithms. To present our approach in precise terms some preliminary notions are needed, in particular the notions of set approximation and
partial dependency of variables in rough set theory.
Let now $B \subseteq C$ be a subset of condition variables which induces the classification $B^*=\{X_1, X_2, \ldots, X_I\}$ of the set of situations contained in E. In
addition, let $M^*=\{Y_1, Y_2, \ldots, Y_J\}$ be a classification of control actions
based on values of control variables in the collection Y, for a set of
situations in E.

Given a class of control actions $Y_j \in M^*$ we can define the lower approximation BY_j of Y_j with respect to the partition $B*$ as

$$BY_j = \{X_i \in B* : X_i \subseteq Y_j\}.$$

The lower approximation or the positive region of the class Y_j specifies an area in the information table within which the control action corresponding to $Des_M(Y_j)$ is functionally determined by conbinations of values of condition variables belonging to the set B. By taking the lower approximation of each class Y_j in M we can obtain the lower approximation of the partition $M*$. That is

$$M_{B*} = \{BY_1, BY_2, \ldots, BY_J\}.$$

In this case, the positive region of this approximation is given by

$$POS_B(M*) = \bigcup_{i=1}^{J} BY_i.$$

The positive region of the approximation of the classification $M*$ by the classification $B*$ is a collection of those situations which can be assigned a unique control action based on values of condition variables contained in the set B. The notion of positive region defined in this way aloows for a simple definition of the degree of functional dependency existing between condition variables B and control variables M. This degree of dependency is of fundamental importance with respect to control algorithm synthesis from examples as it characterizes our ability to produce a control algorithm of given quality (deterministic at the best) from accumulated experience represented in Information Table.

We will say that the set of condition variables B and the set of control variables M are dependent in degree $\gamma(B,M)$ if

$$\gamma(B,M) = |POS_B(M*)| / |E| \quad \text{where}$$

$|\ |$ denotes the number of elements in the set. In other words measure γ is the proportion of those situations in Information Table which can be uniquely assigned correct control action based solely on values of variables from the set B. It can be easily observed that if $\gamma = 1$ then the control variables in M functionally depend on condition variables B and a deterministic control algorithm can be synthesized in this case. Similarly if $\gamma < 1$ then the control algorithm will be nondeterministic with the degree of determinism of this algorithm reflected in the value of γ.

3.3.2. Reduction of Condition Variables

The prime objective of reduction of condition variables is to simplify the control algorithm. The control algorithm obtained by using all condition variables in its decision rules is replaced by an equivalent algorithm (i.e. one which ensures the same accuracy of decisions) with a smaller number of condition variables. Both uniform and non-uniform reduction methods presented here are inspired by the idea of reduct introduced in (3).

The idea of reduct is of fundamental nature in the theory of rough sets. The existence of a reduct different from the set of condition variables implies that some of them can be safely removed from the system without affecting the accuracy of the synthesized control algorithm. Removing some condition variables, in turn, means that less measurements or observations are required to perform given sequence of control actions.

To explain the notion of reduct let us assume once again that $B \subseteq C$ is a subset of condition variables. The subset B is said to be dependent with respect to the set of action (control) variables M if there exists a proper subset $B' \subset B$ such that

$$POS_B'(M^*)=POS_B(M^*);$$

otherwise B is regarded as an independent set with respect to M. Intuitively, the set of variables B is dependent if we can remove some variables from B without affecting the degree of functional dependency between B and M. Formally, in this case

$$\gamma(B',M)=\gamma(B,M).$$

A relative reduct of the set of condition variables will be defined as a maximal (with respect to inclusion relation) independent subset of condition variables.

As it can be verified by an example, in general, there exists more than one reduct. The collection of reducts of C will be denoted by $RED_M(C)$.

Example 5

As it can be verified manually the variables c_1 and c_2 from Table 1 are sufficient to distinguish between control actions 1 and 2. Also, none of them can be removed without affecting this property. Consequently, the relative reduct of Table 1 with respect to the control variable consists of variables c_1 and c_2. As it can be shown this is the only such reduct. Therefore

$$RED_M(C)=\{\{c_1,c_2\}\}.$$

The control algorithm of Example 4 can be greatly simplified by replacing the condition part of decision rules by the reduct discussed in the Example 5. In addition to simplifying the notation we will also connect conditions implying the same action by an OR operator.

IF $((c_1:=1)$ AND $(c_2:=2))$ OR $((c_1:=2)$ AND $(c_2:=6))$ OR $((c_1:=2)$ AND $(c_2:=4))$ OR $((c_1:=2)$ AND $(c_2:=7))$ OR $((c_1:=1)$ AND $(c_2:=5))$

THEN take action 1;

IF $((c_1:=2)$ AND $(c_2:=1))$ OR $((c_1:=3)$ AND $(c_2:=7))$ OR $((c_1:=3)$ AND $(c_2:=1))$

THEN take action 2;

IF $((c_1:=1)$ AND $(c_2:=1))$ OR $((c_1:=2)$ AND $(c_2:=5))$

THEN turn control back to operator.

The reduction of the control algorithm achieved by computing the reduct will be referred to as uniform reduction because the same variables were eliminated from all decision rules. A deeper reduction can be achieved by eliminating some values of the remaining variables. For this purpose a computational procedure as described in (1) can be used. In resulting collection of decision rules some values of condition variables will be replaced by "don't care" symbols.

Due to space limitations this method will not be discussed in this article. Interested readers can find more details in (1). The most comprehensive to date review of the basic results of the deterministic and probabilistic rough set theories can be found in (3).

4. APPLICATIONS AND ALTERNATIVES

Industry wishes to minimize design, installation and operational costs. It, therefore, wants the simplest control system that provides adequate performance (14). Controllers with proportional, integral and derivative gains (PID) are commonly used and as the process dynamics are unknown or are only partically modeled the controller gains are tuned.As processes in industries such as food, paper, cement and petroleum have both variable feed-stocks and high order nonlinear dynamics it may be difficult to tune a PID controller to give good performance under all conditions. A controller that will adapt and change its gains to meet current conditions should be able to give better performance.

Adaptive control, as commonly applied, assumes the controlled process satisfies a dynamic linear equation with uncertain coefficients. The adaptive algorithm estimates the coefficients and uses the resulting estimates in a control law (15). The real system is generally of higher order than the assumed linear model and nonlinear as well (14). Adaptive control may be viewed as a method of finding a near optimal solution to a nonlinear stochastic optimal problem that is too difficult to solve directly (15). The popular adaptive control algorithms are simple and can be implemented in microcomputers (15). Much work has been done on the convergence of various algorithms to stable solutions, (16,17) and in producing adaptive controller designs that will be robust in the presence of disturbances and modeling errors of various forms (18,19). These proofs may give the designer confidence in methods associated with strong results. However, modeling errors and uncertainty are either of unknwon form or are mathematically complex or difficult and have been neglected to make the problem tractable. All adaptive designs must be validated by on line experiment (14,15).

As process controlled by human operators generally have local loop controllers, normally of PID form, these methods of adaptive control, which amount to automatic gain tuning, complement rather than compete with the machine learning approach advocated here. The human operator will adjust process set point, switch units on or off line depending on capacity requirements and take corrective action under failure or alarn conditions. Occasionally he may adjust controller gains, if that function were taken by an automated system it would allow him to concentrate his attentions on his other tasks.

J.H.Burghart and I.Lefkowitz (7) propose a method of automatically adjusting controller set points to compensate for measured disturbances. They assume that the process is well controlled by dynamic regulators and a static input output model model as well as mathematical performance function is available. The control system they propose measures the disturbance statistics and optimizes the function used to compute the set points from the measured disturbances. The attractiveness of this approach lies in attempt to maximize performance. The machine learning approach simply records operator preferences and there is no guarantee that operator preferences are best. Various operators and the management may have differing opinions as to what constitutes good control. This weakness however is also an advantage as mathematical system models and meaningful and general mathematical performance criteria are often not easy to obtain. In addition a human operator is not subject to the restrictions required to make the problem mathematically solvable. The appropriateness of an advisory system over a closed loop optimal system for some situations may be made more clear if we consider the problem of controlling a car. Given the complexity of operating a car so as to stay on the road, avoid other vehicles and compensate for varying weather and road condi-

20. J.C.Francis and R.R.Leitch, Artifact: A Real-Time Shell for Intelligent Feedback Control, in: *Research and Development in Expert Systems*, (M.A.Bramer, ed.) Proceedings of the Fourth Technical Conference of the British Computer Society Specialist Group on Expert Systems, University of Warwick, 18-20 December 1984, pp. 151-162, Cambridge University Press.

21. N.R.Sirada, D.G.Fisher, and A.J.Morris, AI Applications for Process Regulation and Servo Control, *IEE Proceedings*, 134(Pt. D, 4), 251-263 (July, 1987).

22. K.J.Astrom, J.J.Anton, and K.E.Arzen, Expert Control, *Automatica*, 22(3), 277-286 (1986).

23. M.A.Bramer, Expert Systems: The Vision and the Reality in: *Research and Development in Expert Systems*, (M.A.Bramer ed.) Proceedings of the Fourth Technical Conterence the British Computer Society Specialist Group on Expert Systems, University of Warwich, 18-20 December 1984, pp. 1-12, Gambridge University Press.

3. Z.Pawlak, S.K.M.Wong, W.Ziarko, Rough Sets: Probabilistic versus Deterministic Approach, *International Journal of Man-Machine Studies* (to appear).

4. C.J.Harris and S.A.Billings, eds. *Self-Tuning and Adaptive Control: Theory and Applications*, Peter Pereginus Ltd. (1981).

5. K.J.Astrom, Theory and Applications of Adaptive Control — A Survey, *Automatica*, 19, 5, 471-486 (1983).

6. Special Issue on Adaptive Control, *Automatica*, 20 (5), (September, 1984).

7. J.H.Burghart and I.Lefkowitz, A Technique for On-Line Steady-State Optimizing Control, *IEEE Trans on Sys. Sci. and Cyberetics*, SCC-5(2), 125-132 (1969).

8. R.R.Leitch, ed., Special Issue on Artificial Intelligence, *IEEE Proceedings-D: Control Theory and Applications*, 134(4), (July, 1987).

9. Adam Mrozek, Information Systems and Control Algorithms, *Bulletin of The Polish Academy of Sciences (Technical Sciences)*, 33, 3-4, 195-204, (1985).

10. R.S.Michalski, J.G.Carbonell, T.M.Mitchell, Machine Learning, Tioga Publishing Company (1983).

11. S.K.M.Wong, W.Ziarko, INFER - An Adaptive Decision Support System, Proceedings of the 6th International Workshop on Expert Systems and Their Applications, Avignon, France (1986).

12. E.Orlowska, Z.Pawlak, Expressive Power of Knowledge Representation Systems, *International Journal of Man-Machine Studies*, 20, 485-500 (1984).

13. Z.Pawlak, Decision Tables and Decision Algorithms, *Bulletin of the Polish Academy of Sciences*, 33, 9-10, 487-494 (1985).

14. E.H.Bristol, An Industrial Point of View on Control Teaching and Theory, *IEEE Control Systems Magazine*, 6(1), 24-27 (February, 1986).

15. O.L.R.Jacobs, Introduction to Adaptive Control, in: *Self Tuning and Adaptive Control: Theory and Applications* (C.J.Harris and S.A.Billings, eds.), pp. 1-35, Peter Pereginus Ltd. (1981).

16. S.P.Meyn and P.E.Caines, A New Approach to Stochastic Adaptive Control, *IEEE Transactions on Automatic Control*, AC-32(3), 220-226 (March, 1987).

17. G.C.Goodwin and D.Q.Mayne, Parameter Estimation Perspective of Continuous Time Model Reference Adaptive Control, *Automatica*, 23(1), 57-70 (1987).

18. E.Polak, S.E.Salcudean and D.Q.Mayne, Adaptive Control of ARMA Plants Using Worst-Case Design by Semi-Infinite Optimization, *IEEE Transactions on Automatic Control*, AC-32(5), 388-396 (May, 1987).

19. K.S.Narendra and A.M.Annaswamy, A New Adaptive Law for Robust Adaptation Without Persistent Excitation, *IEEE Transactions on Automatic Control*, AC-32(2), 134-145 (February, 1987).

tions, and the poor state of available measurement systems few would propose to place cars under closed loop automatic control. It is found that an advisory system, a driving instructor, available to the new operator will greatly speed the process of learning and is in fact a legal requirement for safety reasons. While complete automatic control is not feasible automatic gain scheduling for changing conditions, automatic transmission, is. Some would prefer a manual transmission with an advisory system to an automatic transmission, however.

As well as the adaptive control methods which are based on presumed underlying mathematical models recent interest in artificial intelligence has led to some experiments with knowledge based controllers which mimic human reasoning (20,21). Astrom (22) argues that an expert system is the best way to organize the heuristic logic that must be incorporated in any practical computer control system for safety and human interaction. This heuristic logic includes such things as checks for equipment operation, alarms, output limits, operating modes, manual to automatic bumpless transfer, and integral term antiwindup logic. Even simple systems contain many rules (22). The acquiring of these rules is the most difficult aspect of any Expert system (23). Consider the difficulty you would have in writing out a rule set for any piece of household equipment you commonly operate. In the context of expert systems the princimple advantage of the machine learning approach is that the rule set is generated completely automatically.

5. SUMMARY

A method of automatically capturing process operators control methods from records of measurements and control actions has been presented. These methods do not require any mathematical model of the process nor do they require any logic engineering to extract a rule base from operational staff.

The operational records are set in an Information Table. Automated methods extract the dependencies between the measurements and control actions and produce the control algorithm as a set of decision rules. The computational methods then reduce these rules to produce the simplest set which are implemented online to control the process or advice the operator.

The mathematical model used here is a rough set approach which divides measurements and control inputs into discrete levels natural to operational staff. This view is independant of continuous variable view taken by adaptive and other learning systems approaches. Adaptive control methods produce reulators which could be used in association with the control algorithms produced here. The algorithms learned from the operators practice function at the process operator level which supervises regulatory control.

REFERENCES

1. Wojciech Ziarko, On Reduction of Knowledge Representation, Department of Computer Science, University of Regina, Regina, Saskatchewan, Canada, S4S 0A2, Technical Report CS-87-04 (April, 1987); Proc. of the Second Symposium on Methodologies for Intelligent Systems, Charlotte, 1987 (Collguia Program).

2. Z.Pawlak, Rough Sets, *International Journal of Computer and Systems Sciences*, 11, 145-172 (1982).

11

EXPERT SYSTEM METHODOLOGY IN PROCESS SUPERVISION AND CONTROL

S.G. Tzafestas*, S. Abu El Ata-Doss** and
G. Papakonstantinou*

* Computer Sci. Div., Natl. Tech. Univ.
Zografou, Athens, Greece

** ADERSA, BP 52, Verrieres Le Buisson
Cedex, France

1. INTRODUCTION

The necessity for applying artificial intelligence (AI) to engineering and control problems originates from the growing complexity of modern engineering and control systems, as well as from the traditional expense, time constraints, and limited availability of human expertise. AI technology offers the tools that enable us to {1-3}:

— capture and retain expertise that was gained over many years of engineering,

— amplify expertise that is needed to successfully deploy new methods and applications, and

— design systems that reason intelligently about nesessasy actions to take in real time, thus freeing operational staff.

The design and application of knowledge-based expert systems (ES) for process fault detection/diagnosis and supervisory control has received a good deal of attention by knowledge engineers, due to the resulting improved efficiency, effectiveness and performance. Expert systems are software programs, supplemented by man-machine interfaces, which use knowledge and reasoning to perform complex tasks at a performance level usually achieved by human experts.

Fault diagnosis and process supervision are knowledge-intensive and experience-based tasks, which in complex processes can sometimes go beyond the capabilities of skilled operators and engineers. Expert systems can provide the critically required assistance for prompt detection and location of process faults, as well as for real-time process supervision.

The main functions of the fault diagnosis and repair task are {4}:

— Symptom interpretation
— Fault diagnosis
— Trouble-shooting
— Repair/replacement
— Test design

— Test/repair scheduling
— Monitoring
— Reasoning with functional models.

After the end of the symptom interpretation and problem diagnosis, the specialist will usually have limited the malfunctions to a few *least repairable units* (LRUs), which are troube-shooted using the specified procedures. When the failed LRU is identified, the appropriate "repair/replacement" action is taken, and the system is retested and its response is monitored to assure that the malfunctioning problem is removed. If the problem cannot be easily diagnosed, the historical dadabase of the system at hand must be consulted in order to obtain a conclusion. If this is not possible, the specialist has to use his knowledge of how the system works (i.e. his *deep knowledge*) to diagnose the problem and, if necessary, to design new tests to test for unusual conditions. Finally the structural and logical coherence of symptoms, tests and repair actions must be checked for final detection decisions, test selection and repair. This last decision level is very often implemented using pattern recognition techniques. The scheduling of tests and repair actions lead to further reduction of the *turn-around time*.

The fault diagnosis/repair task is one of the first tasks to be assigned to ES in process control. A second, equally important, task is the process supervision which involves (besides the fault detection/diagnosis component) the following subtasks:

— process management
— alarm handling
— optimizing control.

A process in operation can be characterized by input and output variables, state variables and process model parameters which define causal relationships between these variables. The process can be subject to different categories of events, namely:

— *system non-stationarities* (e.g. gain variation)
— *environmental non-stationarities* (e.g. perturbation level change)
— *instrumentation non-stationarities* (e.g. actuator or sensor fault).

At the occurence of any event from these categories, the process deviates from its normal functioning known by the operator. The supervisory scheme is not intended to replace the process operator, but has to assist him by providing rapid useful real-time information about the process state. It must be able to detect events or anomalies, to classify them between the different non-stationarity categories, to localize their origin and to evaluate them. Once an event is detected and evaluated, the supervisor must give useful information diagnostics, control strategies, and action planning which have to be analyzed by the operator before taking decisions such as: tolerate the event, change the setpoint, switch to a different operation mode (for example from regulation to tracking), adapt the controller, switch to an emergency control mode or stop the process for security reasons, etc. All these functions represent difficult tasks for even well-trained operators.

Recently, many contributions have been devoted to the development of fault detection and localization procedures for system non-stationarities. Various techniques based on real-time parameter estimation have been presented in Geiger {5}, Geiger and Goedecke {6}, Goedecke {7}, Isermann {8}, and Rault, Jaume and Vergé {9,10}. In Abu el Ata-Doss and Ponty {11} a supervisory control scheme has been proposed for processes which exhibit both system and environmental non-stationarities. This supervision strategy was also based on real-time parameter estimation. In the extensions of this work by Abu el Ata-Doss and Brunet {12,13} a general tool for on-line supervision by an expert system has been developed. Further contributions in expert identification, supervision and control were given by Astrom et. al. {14}, King and Karonis {15}, Moore et al {16}, Moore and Kramer {17} and Sanoff and Wellstead {18}. The development of fuzzy rule-based (linguistic) control of industrial processes can be found in {19-31}. Recent surveys of the work on knowledge-based approach to system diagnosis, supervision and control were given in {32-35}.

Our purpose in this chapter is:

(i) To give the general features of expert systems for process control and supervision (Sec. 2),

(ii) To describe in detail a proposed expert supervisory control scheme following the lines of {12,13} (Sec. 3),

(iii) To provide the fundamentals of fuzzy rule-based control and present an intelligent incremental fuzzy PID controller (Sec. 4), and finally:

(iv) To discuss a number of working expert systems for industrial process supervision and control (Sec. 5).

2. GENERAL EXPERT SYSTEM FEATURES FOR PROCESS CONTROL AND SUPERVISION

2.1. Computer-Aided Control System Design

Expert system tools can be designed and used at both the *control system design* and *process control operation* levels {10-18, 35-37}. The control design task is very much different from other engineering tasks to which the available domain independent expert system tools can be applied. Control system design is a *cut-and-try process* since the control engineer develops tentative plans and then tests them by analyzing the results and executes one or more steps. In the control system design the *initial state* (of the associated AI *planning system*) consists of the given *open-loop system* and a given set of desired *closed-loop performance specifications*. The *final state* consists (mainly) of the *control law* that leads the system to the desired performance (or sufficiently near to it). The most natural data structure for control system design appears to be the *network* or *tree structure,* where each *node* represents a *unique model* of the process and/or a *unique control law. Lines of reasoning* about the design methods are represented by *branches* of the tree. *Children nodes* are generated whenever operations are performed on the contents of a node. Nodes with a common parent share certain common properties.

Control system design needs reasoning at two levels. The *higher level reasoning* produces a general plan for the approach to the design and is based on basic rules about the control system design. The *lower level reasoning* epmloys backward chaining and concerns the mathematical algorithm used for the control design. Although backward chaining leads to more effective control system design, forward chaining (which uses evidential reasoning to generate hypotheses) is more appropriate when the designer wants to have the ability to replan (redesign) the system following failure of a tentative approach. The best results are obtained by a suitable combination of backward and forward reasoning. Very frequently the design of large scale systems is carried out by decomposing the system under control into several smaller sub-systems and employing either hierarchical or decentralized control procedures. For this purpose one needs to accommodate appropriate *recursive functions* in the knowledge-based system.

2.2. Computer-Aided Process Control Operation

We now come at the *operation* of a complex industrial plant which involves three *hierarchical layers* of controls {38}. The *first layer* involves the *control* (usually *PID* or self-tuning control) which is exercised over the process in order to transform the new materials into finished products of given quality. At the *next layers*, the standard regulatory control is supplemented by *sophisticated control* techniques such as *on-line optimization*, *advanced control*, and engineer/operator *information advisories*. The three layers are integrated using a number of *task-oriented* components, distributed over the various parts of the industrial plant and linked via data highways with the control operating console.

Although the utilization of standard digital distributed systems leads to substantial savings (manpower reduction, energy consumption reductions, product yield increase, raw material savings, etc.) work has to be done in order to accomodate facilities for intelligent interpretation of sensor data, fault diagnosis,, etc. The plant status may change dramatically within a few minites and the large number of variables that are monitored and alarms that are provided make necessary the use of expert-system-like reasoning schemes in a way similar to the human expert (operator), who, when confronted with such situations, responds and takes actions within the limited time available. Clearly, in the overall hierarchy of process control, the role of the expert system is above the normal distributed system. The expert system does not replace the distributed control system, but in cooperation with it improves the engineer/operator actions through a better usage of the available information.

2.3. The Real-Time Feature

The expert system of an industrial plant must be able to *scan* and *treat time-varying* (sometimes very rapidly changing) *data*, to *handle interrupts* and *tasks* over priorities, and *initiate actions* depending on current events or according to a schedule. In other words, an industrial complex (e.g. a *cement kiln*, or *refinery*) actually needs an expert system capable of working in *real-time* (i.e. *a real-time expert system*). Two ways for obtaining an expert system working in real-time are the following:

(a) Use of *shared memory* structures for coping with rapidly changing data {39}.

(b) Use of *appropriate interface* between the expert system's inference engine and a data-scanning component (running its own code) and the so called *"focus of attention"* rules {16,40}.

It is useful to mention again here the problems that can arise in a real-time expert system because of the need to *reuse* the *memory space*. To ensure that the expert system will not run out of memory, facts that are no longer used (obsolete) must be eventually removed and the memory space made available for new facts via a *"garbage collection"* process. But if the *firing* of some rules *depends* not only on the presence of certain facts but also on the *absence of others*, garbage collection may lead to *false application of rules* in the knowledge base. One way to avoid this problem is to give the garbage collection task a *higher priority* than other tasks. This however may *"lock"* other tasks out at innapropriate times. The best solution would be to find ways of preventing the characterization of facts as *"obsolete"* by one rule when their absence is taken into consideration in other rules.

An expert control system must be *able to infer* about *transient conditions* of the plant as a whole, about the *time required to initiate* an *action*, and about the *time required* for the *effects of an actions to appear*. One way to achieve such "*temporal reasoning*" is to encode the temporal knowledge as part of specific rules {41}, as for example:

{<*action*>, <*time-delay*>, <*expected result*>}. The main disadvantage of this approach is that the respective information must be known before the real-time situation occurs, and so there is a possibility to use many *ad hoc* rules which finally become *unmanageable*.

To integrate an expert system with a distributed control system an effective communication link between them is needed. The expert system *must have access* to the *plant data base* in as much detail as the human expert has (for both the current and histotical data). A real-time interface must be used whose role is to *access* any *data* available on the *distributed system*, may be *set-points, measurements, high/low limits* or *on/off scan status indicators*, etc. The current process data is used to update the knowledge base.

2.4. Process Models in Expert System Development

We close this section on expert systems features for process control by mentioning that in most cases some kind of process model must be used in the development of expert systems for process control engineers and operators {11,28}. Some models are useful at the *knowledge engineering stage* (to help in the procedure of acquiring human expertise and coding it in a form suitable for the system), and other models are used as *subroutines* by the expert system itself *during execution*. In the fisrt type of models (the so called *functional models*) the knowledge engineer determines the physical quantities (variables) and components of the process and their interrelationships, i.e. the physical paths by which the system components can affect each other. *Functional descriptions from operators* can also be used which together with the previous (*schematic*) ones are used to develop *hypotheses* about the heuristics used by the operators in focusing their attention and organizing their data scanning.
In the second type of models (the so called *procedural models*) process information is encoded into the knowledge base such that to allow the expert system to provide immediate advice of what to do in critical circumstances (*safety actions*).

3. A PROPOSED SUPERVISORY CONTROL SCHEME

3.1. General Supervisory strategy

In this section we propose a general "intelligent" supervisory control scheme which is appropriate for processes subject to both slow and fast changes at known or unknown instants, and also to environmental nonstationarities.

This supervisory scheme has a hierarchical structure with three levels as shown in Fig. 1 {10-13}:

The first level corresponds to the classical *controller-process* loop in which manipulated variables are computed at each sampling point. This level supplies on-line measured input-output variables and/or state variables to the next level.

The second level is the *information generator* element which continuously provides the third level with condensed useful information. It is a

```
                           ┌─ ─ ─ ─ ─►
                           ¦
                           ¦
     ┌─────────────────────────────────────┐
     │ Expert Supervision System            │
     │   - detection                        │
     │   - localization                     │
     │   - classification                   │
     │   - evaluation                       │
     │   - decision making                  │
     │   - action planning                  │
     └─────────────────────────────────────┘
```

                                    condensed
                                    information

     more specific
     information
     extraction

```
     ┌─────────────────────────────────────┐
     │ information generator                 │
     │   - real-time estimation              │
     │   - signal processing                 │
     │   - calculation of various            │
     │     items                             │
     │   - statistical tests results         │
     └─────────────────────────────────────┘
```

     actions to be
     analyzed by   measured
     the operator   variables

```
     ┌─────────────────────────────────────┐
     │       Controller + Process            │
     └─────────────────────────────────────┘
```

Fig. 1 Functional diagram of the supervisory control scheme.

purely analytical layer in which information perception is given
through munerical items and values. A priori information may also be
available in the case of intentional process changes.

The third level is the *expert supervisor* in which both quantitative
and qualitative information must be handled. It is a purely logical element in which the supervisory functions described before are executed.
It must also be able to ask the second level for more perceptive information or adjustment of parameters related to the perception function. Many
actions can be decided simultaneously. The state of functioning of the
process can thus be communicated continuously to the operator, but the
supervision functions and decisions are performed discontinuously as a
human operator would do.

A procedure for implementing the general supervisory strategy in a
simplified example of a turbocharged diesel engine has been presented in
Abu el Ata-Doss and Brunet {12}.

When implementing such a supervisory scheme a fundamental question
arises: how to obtain a highly reliable supervision? The answer would
be: by a carefull integration of appropriate tools and procedures at both
the second and third level of the scheme.

In the following sections we analyze the main characteristics and
difficulties for both levels. This will define the functions for which
techniques and tools have to be chosen.

3.2. The perception level

At the perception level, the main interest is to supply the expert supervision system, i.e. the higher level, with the best possible information about the process state. This information will be given continuously in the form of values of numerical items, judiciously chosen and resulting from calculated and measured variables or sequential statistical tests.

Many questions arise at this level. For example:

— what are the different events which are susceptible to affect the process?

— what variables can be calculated?

— how to determine the signatures of events on these variables?

— what mathematical techniques can be used?

Of course, the answer to all these questions will depend on the particular application treated, but any strategy, which is based on a quantitative (deep) knowledge of the process behavior, will need tools for the following functions:

— modeling
— simulation
— closed-loop adaptive parameter estimation
— signal characterictics analysis
— analytic redundancy,

to which other functions can be added in relation to the particular application.

In order to obtain a reliable percepted information, the items chosen must be sensitive to the different anomalies and the tools used must take into account the operation conditions. This will facilitate the "expert system understanding".

As an example of the problems that are encountered at this level, we consider here the parameter estimation function which plays an important role in the generation of information. The direct identification approach, as described in Figure 2, can be used: the measured inputs and outputs of the process are directly used in the estimation algorithm as in the open-loop case.

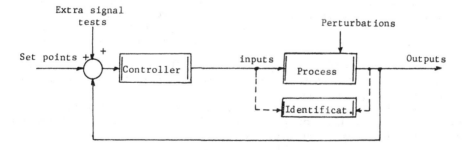

Fig. 2   The direct identification approach in closed-loop operation mode.

The problem with the closed-loop operation mode is that identifiability and accuracy generally arise. In these conditions, estimation procedures and experimental conditions must be carefully chosen. In order to obtain unbiased estimation, a *recursive instrumental variable* type algorithm must be used. An example of such an algorithm is given in Abu el Ata-Doss and Ponty {11}. When set points are constant, identifiability is guaranteed if extra test signals are generated and applied at the process reference input. However this leads to a trade-off between estimation accuracy on one hand and low process perturbation due to the test signals on the other. The supervisor element must take into account the quality of the estimation obtained. An ideal solution to this problem would be to keep a constant level of information in the estimated parameter (i.e. constant parameter uncertainty intervals) by means of suitably chosen test signals.

Another difficulty is due to the trade-off between convergence speed and estimation accuracy, which is common to all adaptive estimation algorithms. This trade-off will affect some functions of the supervision, for example the detection of an event. In detection, it becomes a trade-off between fast but unreliable detections (high rates of false alarms and missed faults) on one hand, and slow but reliable detections on the other.

Concretely, the difficulties in handling the different compromises become difficulties in choosing an tuning parameters, thresholds, time delays, etc., the tuning procedure being organized by the expert level according to the situation.

We close our discussion on the perception level by saying that concerning the tools to be used, there is a large variety suitable for each of the above mentioned functions. Also many sequential tests exist for statistical analysis. The field is very rich and one can easily find what (s)he needs.

3.3. The expert supervisory level

At the expert supervisory level, functions are achieved through organized handling of both *quantitative* and *qualitative* information. The *"knowledge base"* must represent the process behavior in both normal conditions and in typical non-stationary situations. It must be formed by *deep knowledge* using the quantitative model, and *heuristic knowledge* from experience.

An important feature of the supervision is that it must deal with dynamic facts, thus the expert system will possess *temporal properties*, it must have a memory to store recent events together with the operating circumstances related to their occurence. By using stored state (history) information and appropriate *heuristics*, the occurence of premature (inaccurate) decision making can be minimized. A specific strategy to forget the past knowledge and to utilize the information is thus necessary.

Percepted information is supplied on-line by the perception level to the expert supervision element. If the information is judged insufficient for understanding, the expert system must be able to require complementary information from the second level. The essential difficulty at this level is due to the discrimination function. In fact, a given information, e.g. a specific symptom can characterize many different events. In these conditions, more data is needed for *discernability*.

The expert supervision system *reasoning*, from perceptual data, must be essentially based on *discrimination analysis* using both types of logic: *certain* and *fuzzy*. It will operate in *forward* and *backward* inference modes. The rule set will contain two classes of rules:

— *description* rules with certain logic
— *deduction* rules with fuzzy logic.

In the second class, a *probability coefficient* has to be assigned to the rule. This coefficient will represent the degree of confirmation of the proposition based on logical analysis. For rapid decision making, the *interfacing* of the different types of knowledge must be organized appropriately. A cooperative interface with the operator is also of a great importance. The expert system must give the operator:

— *continuously*: a map of the process state
— *occasionally*: the action planning.

On the other hand, the operator can ask the expert system to show him(her) the *logical path* followed in the analysis which yields a particular diagnostic or decision. A graphic tool for visualization of the evolution of some critical variables would be important at this level.

All the above requirements fit into the standard expert system architecture which is shown in Figure 3 for convenience.

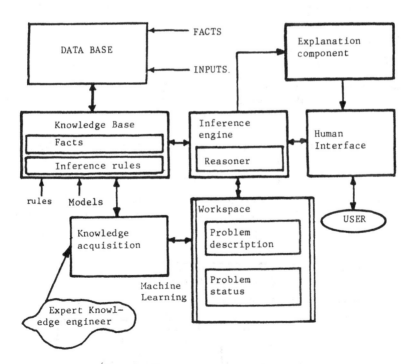

Fig. 3 Expert system architecture.

4. FUZZY RULE-BASED CONTROL

4.1. Basic Fuzzy Controller

The method which is mostly used for the design of rule-based controllers working in a vague and uncertain environment is the so called "fuzzy logic" method developed by Zadeh {42-45}. Fuzzy logic controllers were studied by Mamdani and his co-workers and applied by him and others to various situations such as steam engines, warm water plants, autopilots for ships, cement kilns, robotic arms, water level systems etc. (Assilian and Mamdani {19}, Mamdani {25,26}, King and Mamdani {22}, Mandic, Scharf and Mamdani {27}, Braae and Rutherford {20}, Van Amerongen, Van Nauta Lemke and Vander Veen {30}, Umbers and King {29}, and Larsen {24}).

A schematic representation of a fuzzy control system is shown in Figure 4.

The fuzzy controller is located at the error channel and is composed by a fuzzy algorithm that relates (and converts) significant observed variables to control actions. The fuzzy rules employed depend on the type of the system under control as well as on the heuristic functions used.

The block diagram of the basic SISO fuzzy controlled system developed by Mamdani is shown in Figure 5, where the control u(kT) at time t=kT (T=sampling period) is based on the system error (E) and its rate of change (CE). The control action is expressed linguistically in the form of imprecise conditional statements which form a set of decision rules.

The actual values of the error and the change of error are scaled by the scaling factors (gains) GE nad GC respectively, and the resulting fuzzy variables e(kT) and c(kT) after the quantization procedure (expressed by Q) are given by:

$$e(kT)=Q\{[S-X(kT)] \times GE\}$$
$$c(kT)=Q\{X(kT)-X(k-1)\}^T \times GC$$

These fuzzy variables result in a fuzzy subset U(kT) which, by the decision rule R{·} employed, gives the control element u(kT) required, i.e.

$$u(kT)=R\{U(kT)\}$$

This value is a gain scaled by GU and provides the actual change Δi(kt) in the process input i(kT), i.e.

$$i((k+1)T)-i(kT)=GU \times u(kT)$$

Various forms of relational matrices (or decision tables) R can be found in the literature cited above. Since this matrix usually has very large

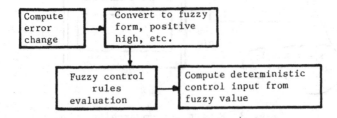

Fig. 4 General representation of a fuzzy control system.

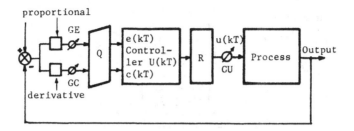

Fig. 5   Basic SISO fuzzy control loop.

dimensionality, simplifications must be sought in order to reduce the computational requirements. The fuzzy controller described above has three scaling factors like the conventional PID controllers.

From the above one can see that the computation of the control action is carried out in four steps, i.e.

Step 1:   Compute the current error and its rate of change.
Step 2:   Translate the error values to fuzzy variables.
Step 3:   Evaluate the decision rules with the aid of Zadeh's compositional rule of inference.
Step 4:   Compute the deterministic input which will be fed to the system under control.

The above steps are included in the block diagram of the fuzzy controller shown in Figure 6 where also the corresponding traditional controller is depicted for the sake of comparison {20}.

4.2. Self-Organized Fuzzy Controller

Procyk and Mamdani {46} have also proposed an improved version of the controller described above, which has more "intelligency" in the sense that it is capable to automatically modify the rules applied according to a measure of deviation of each output from the trajectory $p(kT)$ where

$$p(kT)=\theta\{e(kT),\ c(kT)\}$$

and θ represents the performance decision table used (figure 7).

This controller which was named linguistic (or heuristic fuzzy rule-based) self-organized controller (SOC) was recently applied to control robotic arms (Mandić, Scharf amd Mamdani {27}. The matrix **R**, Figure 7, is given by:

$$\mathbf{R} = T\ \mathbf{J}$$

where T is the sampling period and $\mathbf{J}$ is the system Jacobian matrix which for a system $\dot{\mathbf{x}}=f(\mathbf{x},\mathbf{u})$ is equal to $\mathbf{J}=\partial f/\partial u$. Thus if an output correction $p(kT)$ is needed, then the required correction $\mathbf{r}(kT)$ is found by solving the equation $p(kT)=\mathbf{R}\ \mathbf{r}(kT)$, i.e.

$$\mathbf{r}(kT)=\mathbf{R}^{-1}p(kT)$$

The operation of this controller is now obvious from Figure 7. The performance measure and the increamental model constitute the higher level of the SOC which coordinates the simple fuzzy controller of the lower level. The input correction $\mathbf{r}(kT)$ is fed to the rule modifier which modifies the linguistic rules such that future control actions lead to the appropriate output improvement.

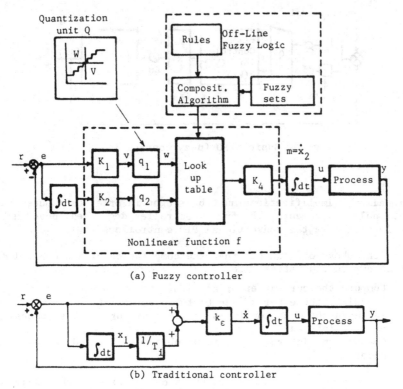

(a) Fuzzy controller

(b) Traditional controller

Fig. 6 Comparison of fuzzy logic (a) and traditional controller (b).

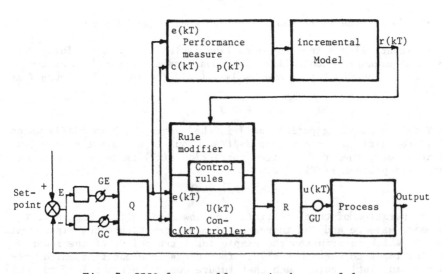

Fig. 7 SISO fuzzy self-organized control loop.

Some theoretical ascpects of the fuzzy controllers of the above type were given by Braae and Rutherford {20} who developed an algebraic model of the controller and using it have studied the loop stability conditions following a purely linguistic approach. They have also demostrated how to synthesize optimal linguistic rules for fuzzy logic controllers.

4.3. Incremental Fuzzy PID Controller

An alternative approach to intelligent fuzzy PID control has been proposed in {47}. This approach assumes that one has available nominal controller parameter settings through some classical tuning technique (Ziegler-Nichols, Kalman, etc.). By using a fuzzy matrix, which is similar to Macvicar-Whelan matrix one can determine small changes on these setting values during the system operation, which lead to improved performance of the transient and steady-state behavior of the closed loop system.

It is assumed that the reader is aware of the PID tuning techniques (analog and digital). For the purposes of the present discussion we shall make use of the Zeigler-Nichols and analytical analog PID tuning techniques, and the digital Kalman tuning technique. A brief review of them can be found in {47}.

Macvicar-Whelan has observed in 1976 that the fuzzy control matrices (i.e. the matrices which contain the fuzzy control rules) of Mamdani and King were incomplete. Due to this, there are situations where there does not exist exact agreement on what value the dependent variable should take for a given value of the independent (error) variable and it derivative.

The fuzzy control matrix proposed by Macvicar-Whelan {48} is shown in Figure 8,and is based on the following principles:

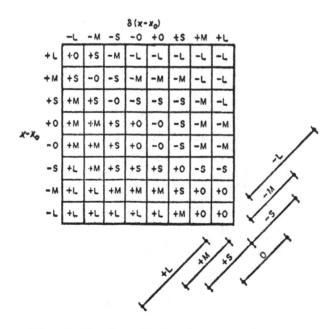

Fig. 8 Macvicar-Whelan fuzzy matrix.

a) If the output has the desired value and the error derivative is zero, then we keep constant the output of the controller.

b) If the output diverges from the desired value, then our action depends on the signum and the value of the error and its derivative. If the conditions are such that the error can be corrected quickly by itself, then we keep the controller output constant or almost constant. Otherwise we change the controller output to achieve satisfactory results.

Most human controllers act by following the philosophy of this matrix. The logic of this matrix approaches the human logic more closely than other fuzzy control mmtrices. This matrix also shows something very fundamental. The designer, prior to designing a fuzzy controller, should study the actions of the human controller, in order to be able to give to the fuzzy controller a more realistic behavior. Note thay if we project the policy of the matrix on an axis perpendicular to the main diagonal, then we will see clearly the form of the policy. One can increase the resolution of the matrix by a factor of 2. This increases the number of the projected parts and reduces the fuzziness of the control policy (Figure 9). The analysis of the projections of the boundaries can be regarded as a study of the variance of the neutral point and the amplitude of the fuzzy fucntion μ_A. Macvicar-Whelan has asserted that using this analysis one can predict the resolution of the projections of the boundaries. An alternative solution would be to start with a small resolution and slowly increase it until we get the desired performance.

This matrix was not used by others, probably because it implements the fuzzy controller in a direct way, and so for each particular case with special requirements it should be varied. Another drawback is the small number of quantization levels used. However, we believe that the logic and structure of this matrix are valuable, and thus we used it for improving existing analog and digital PID tuning procedures as described in the following.

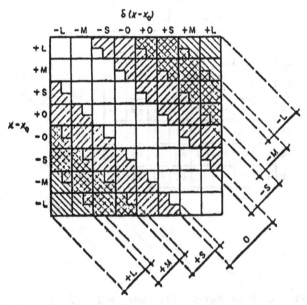

Fig. 9 Increasing the control resolution by a factor of 2.

Our scheme of "*fuzzy tuning*" is based on the parameter values provided by existing tuning techniques (such as the ones discussed in {47}). The idea is to slightly change these values during the system transient so as to improve the characteristics of the step response. The rules used are empirical and are involved in a fuzzy 14x14 control matrix. The problems which must be faced refer to *stability, computation time, cost, quantization,* and *generality* of the approach.

The main issues of our control rules are as follows:

a) Since the integral term is responsible for the overshoot, slightly decreasing it at the moment when the system response exceeds the value 1, one can reduce considerably the overshoot. On the other hand, a small increase of the integral term during the rising of the response, leads to a 10÷20% improvement of the rise time.

b) Since the derivative term is responsible for the flatness of the step response, a small increase of it during rising and in the steady-state eliminates the small oscillations that usually occur.

c) Increasing the proportional term leads to reduced rise time and increased oscillations. Thus finally this term should be decreased to avoid oscillatory behavior.

All these issues are experiential observations of human controllers. Our problem was to accomodate them into a matrix and find the criteria of actually carrying out these changes. The amount of increase/decrease of the proportional, integral and derivative terms is expressed by three coefficients k_1, k_2 and k_3. Our fuzzy PID control matrix is shown in Figure 10. The basis for the control rules is the error $E = X_o - X$ (X_o is the desired and X the actual system output) and its derivative $\Delta E = (X_o - X)$. The sampling of E and ΔE should be made at small time intervals of the order 0.1÷0.2 sec.

ΔE $(X_o - X)$

E	-6	-5	-4	-3	-2	-1	-0	+0	+1	+2	+3	+4	+5	+6
-6	-6	-6	-5	-5	-4	-4	-3	-3	-2	-2	-1	-1	0	0
-5	-6	-5	-5	-4	-4	-3	-3	-2	-2	-1	-1	0	0	0
-4	-5	-5	-4	-4	-3	-3	-2	-2	-1	-1	0	0	0	+1
-3	-5	-4	-4	-3	-3	-2	-2	-1	-1	0	0	0	+1	+1
-2	-4	-4	-3	-3	-2	-2	-1	-1	0	0	0	+1	+1	+2
-1	-4	-3	-3	-2	-2	-1	-1	0	0	0	+1	+1	+2	+2
-0	-3	-3	-2	-2	-1	-1	0	0	0	+1	+1	+2	+2	+3
+0	-3	-2	-2	-1	-1	0	0	0	+1	+1	+2	+2	+3	+3
+1	-2	-2	-1	-1	0	0	0	+1	+1	+2	+2	+3	+3	+4
+2	-2	-1	-1	0	0	0	+1	+1	+2	+2	+3	+3	+4	+4
+3	-1	-1	0	0	0	+1	+1	+2	+2	+3	+3	+4	+4	+5
+4	-1	0	0	0	+1	+1	+2	+2	+3	+3	+4	+4	+5	+5
+5	0	0	0	+1	+1	+2	+2	+3	+3	+4	+4	+5	+5	+6
+6	0	0	+1	+1	+2	+2	+3	+3	+4	+4	+5	+5	+6	+6

Fig. 10 The present fuzzy PID control matrix.

This table involves 14 quantization levels for both E and ΔE. This ensures a closer follow-up of the unit step response.

The linguistic code used is:

+ ⟺ positive change   − ⟺ negative change
6 ⟺ extra large   2 ⟺ small
5 ⟺ large   1 ⟺ extra small
4 ⟺ big   0 ⟺ zero (null)
3 ⟺ medium

The changes of the three terms, which showed good success, are given by the expressions:

$$\left.\begin{array}{l} P=P+CV\{E,\Delta E\}xk_1 \quad \text{(Proportional)} \\ I=I+CV\{E,\Delta E\}xk_2 \quad \text{(Integral)} \\ D=D+CV\{E,\Delta E\}xk_3 \quad \text{(Derivative)} \end{array}\right\} \qquad (1)$$

The parameters k_1, k_2 and k_3 play an important role in the whole procedure, since they determine the range of variation of each term. For example, if some tuning method ensures very small rise time and large overshoot, then the integral term should have a large range of variation, whereas the other terms can remain unchanged. This range of variation should be matched with the stability interval in order to guarrantee stability. Thus the values of k_1, k_2 and k_3 are determined from both the stability analysis and the particular characteristics of the closed-loop response. In general the parameters k_1, k_2 and k_3 provide large flexibility and can be used in conjunction with all available PID tuning algorithms. We have also used the fuzzy control matrix of Figure 11. The only difference with the matrix of Figure 10 is that here the Neperian logarithm of $|CV\{E,\Delta E\}|$ is used. This ensures greater stability around the set point and smoother transition from one value to another. The

ΔE (I_o − I)

E	-6	-5	-4	-3	-2	-1	-0	+0	+1	+2	+3	+4	+5	+6
-6	-1.79	-1.79	-1.60	-1.60	-1.38	-1.38	-1.10	-1.10	-0.69	-0.69	0.00	0.00	0.00	0.00
-5	-1.79	-1.60	-1.60	-1.38	-1.38	-1.10	-1.10	-0.69	-0.69	0.00	0.00	0.00	0.00	0.00
-4	-1.60	-1.60	-1.38	-1.38	-1.10	-1.10	-0.69	-0.69	0.00	0.00	0.00	0.00	0.00	0.00
-3	-1.60	-1.38	-1.38	-1.10	-1.10	-0.69	-0.69	0.00	0.00	0.00	0.00	0.00	0.00	0.00
-2	-1.38	-1.38	-1.10	-1.10	-0.69	-0.69	0.00	0.00	0.00	0.00	0.00	0.00	0.00	+0.69
-1	-1.38	-1.10	-1.10	-0.69	-0.69	0.00	0.00	0.00	0.00	0.00	0.00	0.00	+0.69	+0.69
-0	-1.10	-1.10	-0.69	-0.69	0.00	0.00	0.00	0.00	0.00	0.00	0.00	+0.69	+0.69	+1.10
+0	-1.10	-0.69	-0.69	0.00	0.00	0.00	0.00	0.00	0.00	0.00	+0.69	+0.69	+1.10	+1.10
+1	-0.69	-0.69	0.00	0.00	0.00	0.00	0.00	0.00	0.00	+0.69	+0.69	+1.10	+1.10	+1.38
+2	-0.69	0.00	0.00	0.00	0.00	0.00	0.00	0.00	+0.69	+0.69	+1.10	+1.10	+1.38	+1.38
+3	0.00	0.00	0.00	0.00	0.00	0.00	0.00	+0.69	+0.69	+1.10	+1.10	+1.38	+1.38	+1.60
+4	0.00	0.00	0.00	0.00	0.00	0.00	+0.69	+1.10	+1.10	+1.38	+1.38	+1.60	+1.60	
+5	0.00	0.00	0.00	0.00	0.00	+0.69	+0.69	+1.10	+1.10	+1.38	+1.38	+1.60	+1.60	+1.79
+6	0.00	0.00	0.00	0.00	+0.69	+0.69	+1.10	+1.10	+1.38	+1.38	+1.60	+1.60	+1.79	+1.79

Fig. 11 An alternative fuzzy PID control matrix.

results obtained using this table are also satisfactory, but the values of k_1, k_2, k_3 are found to be larger.

The stability can be tested by various techniques. In our studies we used Routh's table in the continuous-time case and the corresponding table, obtained through the bilinear transformation $z=(w+1)/(w-1)$, in the discrete-time case.

The most serious problem faced in our experiments was that of quantization. The right quantization of E and ΔE ensures a good evaluation of the controller performance. We found the logarithmic quantization to be the most successful. We have used the following quantization levels, $Q_{max}=0.02$, $DQ_{max}=0.02$. Let us see in detail the logarithmic quantization. If the error E is greater than $Q_{max}/2$, then it is quantized in a value between +1 and +6. If $E<-Q_{max}/2$, then it is quantized in the interval $[-6, -1]$. If $0<E<Q_{max}/2$, then it is quantized to 0. If $-Q_{max}/2<E<0$, then it is quantized to -0. Analogous quantization is applied to ΔE. It should be noted that these quantization levels are sufficiently large to cover the presence of noise. However this quantization was behaved quite well.

5. EXAMPLES OF PROCESS SUPERVISION AND CONTROL EXPERT SYSTEMS

In this section we briefly discuss a representative set of existing expert systems for industrial process supervisory control.

Example 1: PID/Self-tuning Expert Control

The principal operations (functions) of *PID/Self-tuning* expert controller are the following {14}:

— *Back-up control*: PID control designed by *Ziegler-Nichols* methods. An estimator of the critical gain K_c and critical period T_c is needed. Here the control signal u(t) is calculated by

$$u(t)=K\{e(t)+\frac{1}{T_i}\int_0^t e(\tau)d\tau+T_d\frac{de(t)}{dt}\}$$

where e(t) is the error signal.

— *Estimation*: Parameter estimation, estimation supervisor, excitation supervisor, perturbation signal generator, and jump detector.

— *Self-Tuning Control*: Use of some self-tuning regulation method.

— *Learning*: Get regulator parameters, smooth-and-store regulator parameters, and test scheduling hypothesis.

— *Main monitor*: Stability and robustness supervisor (computation of means and variances).

A prototype system based on these functions was build using the *LISP* and *PASCAL* facilities of a *VAX 11/780* running under *VMS*. The software package *EUNICE* was used to create a *UNIX* environment under VMS. The expert system was based on the *OPS4* expert system shell.

Example 2: Process Intelligent Control

An expert system for process supervision and control, commercially available, is the so-called *PICON* (*Process Intelligent Control*) which has been developed by *LISP Machine Inc* (*LMI*) in USA {40,41,49}. The *PICON* expert system operates on a *LISP* machine interfaced with a conventional distributed control system. Up to 20,000 measurement points and alarms can be assessed. The real-time data interface is via an integral multibus connected through an interface board to the distributed process control system. The distributed system does not transmit all measurements and alarms in a mixed scan way, but rather the process data are accessed as required for inference. This is somehow similar to a human expert operator action, who gives more attention or scans the process operation selectively, using his expertise. The system mimics a human operator in invoking rules and procedures for safety, fault diagnosis supervision, or other purpose when required and involves assembling information and primary analysis in order to allow inference about the process under consideration.

For computational efficiency two parallel processors with a shared memory are used. One of the processors is a *68010* programmed in *C* code and the other is the *LISP* processor. The expert system package is designed in such a way that an algorithm of arbitrary structure can be dynamically loaded into the *68010 processor* from the *LISP processor*. The expert system has also the ability to "*focus attention*" to a specific area of the process plant, and put all associated measurements and rules for that area on frequent scan. This can be performed under the control of the *LISP program*. The structure of the *LMI* hardware system is as shown in Figure 12, whereas the overall structure of the *PICON* expert system is as shown in Figure 13.

The *rules* are designed so as to combine *process measurements* and *process dynamic behaviour*, as shown, for example, in the following simple tank level rule:

```
IF    tank-level is-greater-than 80%
OR    tank-level is-greater-than 60%
AND   rate-of-increase tank-level is-greater-than 10%
THEN  tank-level is too high.
```

The *knowledge-base* is organized using the *deep knowledge* approach (the underlying plant structure is entered into the knowledge base by graphical construction and by specifying attributes) and each item of the schematic has an associated knowledge frame. The schematic capture of process knowledge is actually based on an interactive procedure.

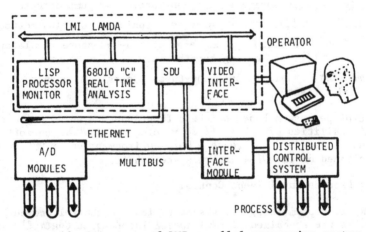

Fig. 12 Structure of LMI parallel processing system.

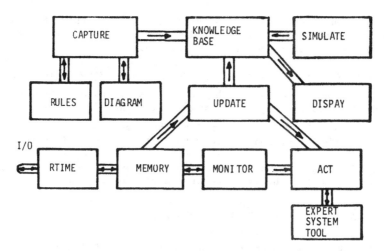

Fig. 13 Overall structure of the Process Intelligent
Control Expert System.

The *heuristics* are represented by *frames of knowledge* having dynamic
attributes and including slots for rules, attributes related to the rules,
particular process units or problem types, attributes related to the use
of knowledge, records of knowledge, authors and editors, etc.

The *inference engine* is designed so as to gather data, derive conclusions, make decisions on future actions, and implement them in a continuously changing environment. *Measured* or *inferred values* are *time-
stamped*, i.e. one can know how old they are. Every variable has a *currency
interval*, so that state information older than the currency interval is
not used, but the state is re-evaluated.

PICON has an *event scheduler* which is capable of initiating specific
actions at specific times in future, and helps the inference engine in
reminding itself to do the proper things. Another facility of PICON is
the *cycle system* which controls the initiation of activities on a regular
basis, in contrast to the event scheduler which is used for scheduling
single events.

Using *PICON* several different process monitoring activities can take
place without interfering with each other. The *concept of time* is ab-
stracted, and thus the system can work either in real or in simulated
time. Finally, several tools of *knowledge retrieval and testing* have been
incorporated in an assembled form in *PICON*. These include a natural language menu and a simulation facility.

Example 3: A Rule-Based Self-Tuning Supervisor

The rule-based self-tuning supervisor proposed in {50,51} is essentially of the type described in Section 3. The self-tuned control system has two levels. The lower level is the standard feedback control
loop with the process and the controller. The second level is the adaptation mechanism which adjusts the controller parameters {52}. A self-
tuning controller works satisfactorily if all preconditions are satisfied.
But in practice, the operation preconditions may be violated and appropriate control parameters and other "a priori" factors may not be known. The
required tuning can be undertaken by a supervisor whose functions include
tracing-back of preconditions violations, parameters estimation testing,

and modification of controller design factors. Thus a third feedback level is needed to supervise and coordinate the self-tuning controller through the appropriate supervisory functions.

The following two supervisory functions were adopted in {50}:

— *The monitoring function,* which involves the calculations based on values provided by the self-tuning levels 1 and 2.

— *The decision function,* which employs the numerical data provided by the monitoring function in order to decide which actions to take.

The monitoring procedure used in {50} is based on the following recursive expression for the variance of the monitored values:

$$s^2(t) = \frac{1}{N-1} \sum_{i=1}^{N} \{x_i(t) - \bar{x}(t)\}^2 \quad (N=20) \tag{2}$$

where $\bar{x}(t)$ is the mean value of $x(t)$. Regarding the testing of the estimated parameters the following quantity was used

$$dp = \sum_{i=1}^{n} |p_i - \bar{p}_i| \tag{3}$$

where $\bar{p}_i$ is the mean value of the ith component p_i of the estimated parameter vector. This is a modification of the test quantity used in {11}. Other numerical quantities suitable for monitoring purposes are the trace of the information matrix, the error signal, the prediction error and the trace of the covariance matrix.

The decision function can lead to several different actions, e.g. modifications of parameters (such as the system order, time delay and sampling period), selection of various algorithms (e.g. recursive least squares estimator etc.), and switch on/off different algorithms (e.g. the controller parameter adjusting algorithm when estimator burst is detected). The decision function may be regarded as a refined logical procedure mainly implemented by simgle *IF-THEN-ELSE* instructions comparing numerical values with the appropriate thresholds. The structure organization of the rule based supervisor and the expert supervision level adopted in {50} is shown in Figures 14 and 15.

The *representation translator* determines the linguistic equivalent of the data base (specification table). The *specification* table contains the "a priori" expert knowledge of the process (maximum and minimum thresholds, maximum and minimum variances, etc). The *knowledge transformer* receives a dynamic list representation of the data base from linguistic equivalent, and prepares the data required by the inference engine to draw conclusions. The *representation translator* is made of two components the *initial representation translator* which only transforms the numerical values of evidences, and the *facts representation translator* which transforms the monitoring level variables (calculated at every sampling instant) and must be executed continuously.

Rule examples are the following:

Definition Rule:

    IF   MATRIX H TRACE LOW *AND*
            CONTROL VARIABLE LOW *AND*
            SYSTEM VARIANCE LOW

    THEN PROCESS IS NOT PERTURBED
    END-RULE

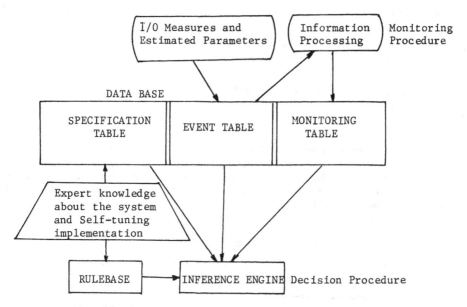

Fig. 14   Organization of the rule-based supervisor.

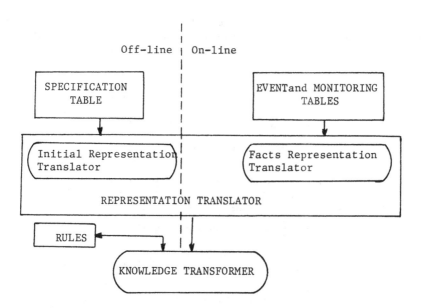

Fig. 15   General structure of the expert supervision level.

Decision Rule:

```
IF     PROCESS IS NOT PERTURBED AND
       PREDICTION ERROR LOW AND
       COVARIANCE MATRIX TRACE HIGH

THEN
       USE FIXED FORGETTING FACTOR

END-RULE
```

Definition rules are used to establish a property or state of the system under control, and decision rules to decide which procedures to invoke (e.g. variable modification, activation of algorithms, etc). At each decision layer, the rules are structured as a dynamic list of elements with two pointers to a respective dynamic list of antecedents and consequents. Reasoning is carried out in the rule order. A chain rule mechanism is used for the case where more than one level must be executed.

The first two supervision functions implemented are:

(i)   Switching among the following states (see Table 1):

— Start-up procedure
— Closed-loop self-tuning control
— Back-up control
— Closed-loop fixed minimum variance control

(ii) Covariance matrix up-dating method selection among the following methods:

— Fixed forgetting factor
— Exponential forgetting factor
— Variable memory length forgetting factor
— Addition to the covariance matrix
— Reset to the covariance matrix.

Example 4: An Expert System for Distillation Control Synthesis

In {53} a general knowledge-engineering approach to the synthesis of process control systems was presented and applied to a distillation system. Upon the assumption that the basic design of the process system has already been finished, this procedure is summarized as in Figure 16.

S.1 DEFINITION OF A PROCESS SYSTEM
↓
S.2 DETERMINATION OF CONTROL OBJECTIVES
↓
S.3 SYNTHESIS AND SELECTION OF POSSIBLE
CONTROL LOOPS IN EACH UNIT
↓
S.4 ANALYSIS OF CONTROL LOOPS IN EACH UNIT
AND COORDINATION AMONG CONTROL LOOPS
IN THE PROCESS SYSTEM
↓
S.5 DETAILED DESIGN OF EACH CONTROL LOOP
↓
S.6 CONFIRMATION OF CONTROL LOOP PERFORMANCE
BY USING PROCESS DYNAMIC SIMULATORS
↓
S.7 CONFIRMATION AND ADJUSTMENT OF CONTROL LOOPS
IN REAL PLANTS

Fig. 16 General procedure for control system design.

Table 1. State selection supervision rules

State	Condition	Action
Start-up procedure	Acceptable parameter estimation	Activate self-tuning control
Self-tuning control	Good param. estim. Good control behavior	Fixed minimum variance control
	Bad control behavior	Activate back-up control
Back-up control	Acceptable param.estim. Good control behavior	Activate self-tuning control
Minimum variance control	Changes in param. estim. Poor control behavior	Activate self-tuning control
	Bad control performance	Activate back-up control

Experimental results have been reported in {50} concerning a 2-time constant system.

If the flow rate of the feed F is specified to be constant, the five manipulated variables in a basic distillation system (Figure 17) are:

C: cooling medium flow rate
R: reflux flow rate
D: distillate flow rate
H: heating medium flow rate
B: bottom flow rate.

If C is not specified to be one of the manipulated variables one can use the flow rates of the overhead vapor V and the condensed liquid E.

Among the numerous controlled variables of the distillation system, five variables can be selected as controlled variables for the primary control loops, and three of these five must be the column overhead pressure P, and the liquid levels of the overhead drum LD and the bottom LB.

The basic inference procedure of the expert system employed has the form shown in Figure 18.

The definition of the distillation system and the determination of the control variables are included in the first five contexts. If any conflict is detected by the cheking contexts involved, inference control is given to the previous context. The "CONTROL OBJECTIVES" context contains appropriate production rules for selecting control variables. The contexts from "FIND LOOPS" through "CHECK THE RESULT" are used to synthesize the control loops for the distillation system.

The control system synthesis strategies are:

— The modified position ordered method is used to resolve conflicts between production rules.

— The production rules are assigned priorities. The rule with the highest priority should be fired first. Dual composition control and single composition control are assigned higher priorities.

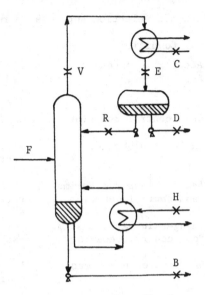

Fig. 17 Schematic of a basic distillation system showing the manipulated variables.

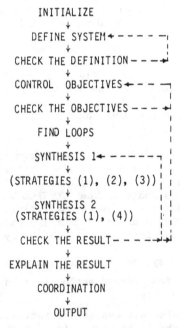

Fig. 18 Inference procedure of the control synthesis expert system.

— A control variable having only one control loop available must be
    selected.

— A control variable having more than one control loops available
    should be treated later via the heuristic knowledge obtained during
    the control system synthesis.

The results of the synthesis are evaluated in context "CHECK THE
RESULT". If no complete pairing-up exists between manipulated and controlled variables, the inference control is given to the upper contexts
as shown in Figure 18.

The coordination between cooperating distillations is carried out in
the context "COORDINATION".

The knowledge base contains 95 production rules, half of which are
related to heuristic knowledge and the rest to utility functions (input,
output, context control, job initiation and symbolic manipulation). The
data base consists of entities, attributes and values. The nine entities
involved are classified into four groups namely: inference control, variables in control loops, system description, and control loops.

The expert system was employed to synthesize a set of control loops
in a refinery's debutanizer (see Table 2).

A schematic of the five synthesized control loops is shown in Figure
19. Normally, P is controlled by C and occasionally by DV. This is included as a secondary loop.

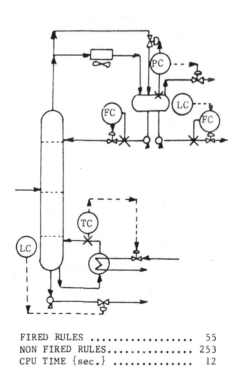

```
FIRED RULES .................. 55
NON FIRED RULES.............. 253
CPU TIME {sec.} ............. 12
```

Fig. 19 Final control loops synthesized by the inference.

Table 2. Description of a refinery's debutanizer

Entity	Basic item	Specification
EQUIP	Feed	Depropanized gasoline
	Feed from	Depropanizer bottom
	Distillate	Butane
	Distillate to	Storage tank
	Off gas	Occasionally
	Off gas to	Flare stack
	Bottom	Gasoline
	Bottom to	Storage tank
	Bottom flow rate	Medium (not small)
	Cooling medium	Water
	Heating medium	Heating oil
VAR	Manipulated variables	C, D, R, B, H, DV
	Controller variables	
	Basic	P, LD, LB
	Specified	R, TS

Example 5: Expert Supervisory Control of a Turbo-charged Diesel Engine

The supervisory control scheme proposed in Section 3 was applied to an example of turbo-charged diesel engine {12}. In such a system, expert supervision would optimize the engine functioning and its management by non qualified staff, increase the availability through a better knowledge of the engine health status, and allow predictive actions of maintenance for objective realization.

Figure 20 shows a part of the process with:

— a sigle input, the manipulated variable X representing the rack position.

— two outputs, namely the controlled engine speed Y1 and the press-ure output Y2 from the compressor.

The expert system receives real-time information through the follow-ing items provided by estimation, correlation computation and residual or tests:

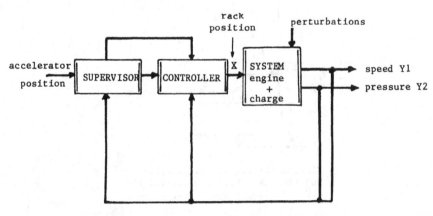

Fig. 20   Block diagram of the expert supervisory control system.

— The estimated gains of both transfers $X \to Y1$, $X \to Y2$.

— The mean values of these gains.

— The difference between Y1 and its reference value (or set point).

— The difference between Y1 and Y2 and their corresponding values determined from model estimation.

— The residuals from analytical redundancy relationships.

— The variances of some of the above differences and residuals.

— The ratio between the estimated slope of the autocorrelation function of X and its nominal value.

— The sequential probability ratio test (SPRT) {9} and threshold test results on some items.

For simplicity, it was assumed that the nonstationarities which can affect the various elements of the process are limited to the following:

— *Actuator* (X):   blockage, efficiency loss, and noise variation.

— *Sensor* (Y2):   blockage and noise variation.

— *Engine/turbine*:   Degradation.

— *Compressor*:   Degradation, pumping.

A diagnostic component has been included with both description and production rules of the form:

Description rule

BLOCK-Y2 *AND* NOISE-Y2 *CONSTITUTE* FAULT-Y2

Production rule

RESIDUAL-REF-OUTPUT Y1 *IS IN FAVOR*

BLOCK-X < -50/1., -1./0., 1./0., 50./1. > ;

The sequence at the end of this rule represents a function which affects the certainty factor CFx100% in the associated deduction (see Figure 21).

The supervisory control is implemented by the expert system *SAM* (Système d'Aide en Médecine) which was originally designed for medical diagnosis. SAM is suitable to implement the supervisory functions, since it is mainly oriented towards hypothesis discrimination. It allows asking for complementary tests, making diagnostic decisions, and taking pronouncing repair (therapeutical) actions. SAM is written in FORTRAN (about 10,000 instructions), uses certain and fuzzy logic, operates in both forward and backward inference modes, and allows rapid execution through a pre-queuing of rules, each rule being inferred once.

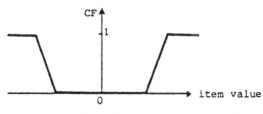

Fig. 21

Several nonstationarities have been tested, sometimes simultaneously, for which supervisor diagnostics in real-time was applied. The results showed that the proposed supervision strategy has a high performance.

Example 6: Simulated Results of the Incremental Fuzzy PID Controller

Here we shall present some representative simulation results of our incremental fuzzy PID controller based on the Ziegler-Nichols and Kalman tuning techniques {47}.

(a) Fuzzy Ziegler Nichols Techniques

The fuzzy matrix of Figure 19 was applied to a PID controller tuned through the Ziegler-Nichols techniques. The system under control has three time constants T_1, T_2 and T_3 i.e.

$$G_2(s)=K/\{(1+sT_1)(1+sT_2)(1+sT_3)\}$$

To determine T_x and K_c we remove the integral and derivative terms. The resulting characteristic polynomial is: $A_1 s^3+A_2 s^2+A_3 s+A_4$ with: $A_1=T_1 T_2 T_3$, $A_2=T_1 T_2+T_1 T_3+T_2 T_3$, $A_3=T_1+T_2+T_3$ and $A_4=1+KK_p$. From Routh's table we find the values:

$$K_c = \frac{A_2 A_3/A_1 -1}{K}, \quad T_x=2\pi\left[\frac{A_2}{1+KK_c}\right]^{\frac{1}{2}}$$

Thus we set

$$K_p=0.6K_c, \quad \tau_i=T_x/2, \quad \tau_d=\tau_i/4$$

The resulting closed-loop system was simulated with the 4th-order Runge-Kutta technique {dashed curves in Figure 22 (a-c)}.

Then we used the fuzzy matrix of Figure 11, with the quantization of E and ΔE being made every 0.08 sec. To assure stability of the closed-loop system we again used Routh's table to the closed loop polynomial

$s^4+b_1 s^3+b_2 s^2+b_3 s+b_4$. The conditions for stability are

$$b_1 b_2>b_3, \quad b_3 L_1>b_1 L_2, \quad b_1=A_2/A_1, \quad b_2=(A_3+KK_d)/A_1$$

$$b_3=(1+KK_p)/A_1, \quad L_1=b_2-(b_3/b_1), \quad L_2=b_4$$

A Pascal subroutine was developed that implements this stability analysis.

Some actual experimental results obtained using the policy (1) are shown in Figure 22 (1-c). Figure 22a refers to $T_1=0.1$ sec, $T_2=0.2$ sec and $T_3=0.7$ sec. The overshoot from 15% is reduced to 1% and the rise time from 0.175 sec to 0.1 sec ($k_1=0.5$, $k_2=0.2$, $k_3=0.4$). Figure 22b corresponds to $T_1=0.6$, $T_2=0.7$ sec and $T_3=0.8$ sec. The overshoot is reduced from 14% to 2% and the rise time from 0.6 sec to 0.55 sec ($k_1=0.2$, $k_2=0.27$, $k_3=0.9$). Finally in Figure 22c the overshoot is reduced from 17% to nearly 0% and the rise time from 0.5 sec to 0.3 sec ($k_1=0.4$, $k_2=0.4$, $k_3=0.4$).

(b) Fuzzy Digital Controller Tuning Based on Kalman's Technique

The fuzzy correction matrix of Figure 10 was applied to a digital PID controller tuned by Kalman's technique. A system $G_2(s)$ with two time

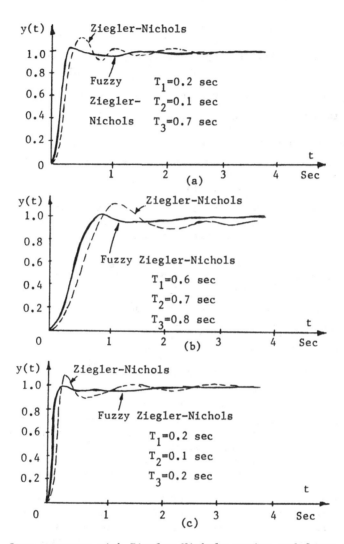

Fig. 22 Step response with Ziegler-Nichols tuning and fuzzy control.

constants T_1 and T_2 was assumed with D.C. gain K. The combination of $G_2(s)$ with the zero-order hold gives the pulse transfer function H(z) to be controlled. The digital PID controller has the form:

$$G_1(z)=K_p\{1+(T/\tau_i)/(1-z^{-1})+(\tau_d/T)(1-z^{-1})\}$$

$$K_p=A_1/(1+\gamma)A_2A_3K, \quad \tau_i=TA_1/A_2A_3, \quad \tau_d=T/A_1$$

where

$$\gamma=C4/(C3+C4), \quad A_1=e^{T/T_1}+e^{T/T_2}-2, \quad A_2=e^{T/T_1}-1, \quad A_3=e^{T/T_2}-1$$

The system was simulated by directly converting $G_1(z)$ and H(z) in difference equation form. The values of τ_i and τ_d were chosen as in (10), The value of $1+\gamma$, needed in the calculation of K_p in (10), was taken equal to 2 in order not to have a large overshoot.

The correction matrix was applied with

$$K_p=K_p+CV\{E, \Delta E\}xk_1, \quad KC_i=KC_i+CV\{E, \Delta E\}xk_i \quad (i=1,2)$$

where $KC_1=K_p(T/\tau_i)$ and $KC_2=K_p(\tau_d/T)$.
The value of T was selected (in each case) such that to satisfy all requirements mentioned in the text. In all but one case it was taken to be less than half the smallest system time constant.

The results in some cases are shown in Figure 23 (a,b). In figure 23a the Kalman method shows a 4% overshoot and a 0.2 sec rise time. Our method shows almost zero overshoot and 0.15 sec rise time ($k_1=0.02$, $k_2=0.017$, $k_3=0.282$). The corresponding values shown in Figure 23b are:

Kalman method : h=3%, t_{rise}=0.4 sec

Present method: h=0%, t_{rise}=0.35 sec

for a system with T_1=0.4 sec, T_2=0.6 sec and T=0.2 sec (k_1=0.01, k_2=0.012, k_3=0.1).

This PID controller is of the *"intelligent"* type since the changes to the parameter settings are made in a fashion similar to the human operator action. It is actually a controller of a higher hierarchical level than the classical analog or digital PID controllers, and prossesses in an accumulated form all the relative human experience. One of the main characteristics of this controller is its small computational requirement. The implementation on a digital PID controller needs 3 multiplications, 6 divisions, 7 additions and 1 substraction. The total implementation time on a Z-80 processor is 959 μsec (i.e. about 1 msec). Analogous computation time is required for the implementation of the method on an analog PID controller. With a 16-bit microprocessor the computational times are reduced to 10% of the above figure. The substantial improvement of the achieved performance, in combination with the above small computational time, makes the method ideal. The additional memory required for the a priori storage of the fuzzy matrix is very small. The 1 msec computational time is about 10% of the computational times needed for the SOC controller of Mamdani. Our method does not propose the replacement of the existing PID controllers. One has merely to complement them with a "chip" which improves their performance. The additional cost is very small for the improvement abtained. For example, using the Ziegler-Nichols controller an overshoot of the order of 10-15% is obtained which is undesirable in chemical processes. The same controller with our improvement yields an

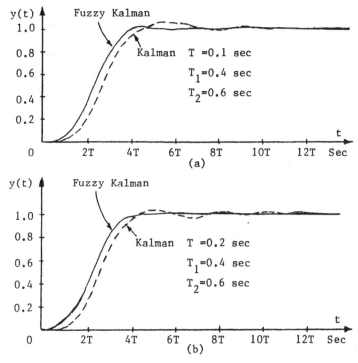

Fig. 23   Step response with digital PID controller tuned by Kalman's technique and fuzzy correction.

overshoot of $1^{o}/oo$. Our controller behaves very well even in the presence of noise, since, if a sudden large increase of the error occurs, the coefficients k_1, k_2, k_3 are varied so as to minimize the error. Of course this depends on the type of quantization. In our experiments we always obtained good results with logarithmic quantization.

6. CONCLUSION

In this chapter we presented an overview of the application of the knowledge-based expert system methodology in the area of process super-vision and control. Although it is very difficult to discuss all aspects, techniques and practical issues in a single chapter, an effort was made to include the most fundamental points.

Thus both the general expert system properties which must be pos-sessed by a knowledge-based intelligent supervisor were discussed, and a representative set of existing expert supervision/control systems were presented. In addition to this, a detailed description of a general expert supervisory control scheme, as well as an outline of the state of-art of intelligent fuzzy control were presented.

The supervisory control scheme proposed here has a hierarchical structure with three-levels; the controller-process level, the perception or information generator level which is purely analytical (techniques for modelling, simulation, closed-loop parameter estimation, signal processing and analytic redundancy are used at this level), and the expert super-vision system level which has a purely logical nature. This third level uses the accumulated information continuously provided, via numerical items values by the perception level, to detect, localize and classify events, to evaluate their impact, to make decisions and initiate appropriate actions. The difficulties encountered at the second and third level of this scheme have been discussed. A large variety of tools can be used depending on each particular situation. Such tools and techniques can be found in this book and in {54-60}. The field is rapidly expanding both methodologically and practically, and many important results and applications are expected to appear in the very near future.

REFERENCES

1. W.S. Faught, Applications of AI in Engineering, *Computer*, 17-27 (July, 1986).

2. M. Gerencser and R. Smetek, Artificial Intelligence: Technology and Applications, *Miltech: Military Technology*, (6):67-76 (1985).

3. R.A. Herrod and B. Papas, Artificial Intelligence Moves into Industrial and Process Control, *The Industr. Process Control Magaz.*, 45-52 (March, 1985).

4. L.F. Pau, Survey of Expert Systems for Fault Detection, Test Generation and Maintenance, *Expert Systems*, 3(2):100-111 (1986).

5. G. Geiger, Monitoring of an Electrical Driven Pump Using Continuous-Time Parameter Estimation Methods, *Proc. 6th IFAC Symp. on Identification and Syst. Param. Estimation*, Washington, U.S.A. (June, 1982).

6. G. Geiger and W. Goedecke, Fault Detection by Modeling, Parameter Estimation and Statistical Tests, *Proc. 4th IMEKO Symp. on Measurement Theory - TC7*, Bressanore, Italy (1984).

7. W. Goedecke, Fault detection in a Tubular Heat Exchanger Based on Modeling and Parameter Estimation, *Proc. 7th IFAC Symp. on Identification and Syst. Param. Estimation*, York, England (July, 1985).

8. R. Isermann, Process Fault Detection Based on Modeling and Estimation Methods, *Proc. 6th IFAC Symp. on Identification and Syst. Param. Estimation*, Washington USA (June, 1982).

9. A. Rault, D. Jaume and M. Vergé, Industrial Processes Fault Detection and Localization, *Proc. 9th IFAC World Congress*, Budapest, Hungary (1984).

10. A. Rault, D. Jaume and M. Vergé, On-line Identification and Fault Detection of Industrial Processes, *Proc. 6th IFAC Symp. on Identification and Syst. Param. Estimation*, Washington, USA (June, 1982).

11. S. Abu el Ata-Doss and P. Ponty, Supervision of Controlled Processes in Non-stationary Conditions, *Proc. 7th IFAC Symp. on Identification and Syst. Param. Estimation York*, England (July, 1985).

12. S. Abu el Ata-Doss and J. Brunet, On-line Expert Supervision for Process Control, *Proc. 25th CDC*, Athens, Greece (Dec. 1986).

13. S. Abu el Ata-Doss and J. Brunet, Conception of Real-Time Expert Supervision Based on Quantitative Model Simulation of the Process, *1st ICIAM*, Paris, France (June, 1987).

14. K.J. Aström, J.J. Anton and R.E. Arzen, Expert Control, *Automatica*, 22(3), (1986).

15. R.E. King and F.C. Karonis, Rule-Based Systems in the Process Industry, *Proc. 25th CDC*, Athens, Greece (Dec. 1986).

16. R.L. Moore, L.B. Hawkinson,. C.G. Knickerbocker and L.M. Churchman: A Real-Time Expert System for Process Control, *Proc. 1st Conf. on Artificial Intelligence Applications*, Denver, U.S.A. (Dec., 1984).

17. R.L. Moore and M.A. Kramer, Expert Systems in On-line Process Control, *Proc. ASILOMAR*, U.S.A. (Jan., 1986).

18. S.P. Sanoff and P.E. Wellstead, Expert Identification and Control, *Proc. 7th IFAC Symp. on Identification and Param. Estimation*, York, England (July, 1985).

19. S. Assilian and E. Mamdani, An Experiment in Linguistic Synthesis with Fuzzy Logic Controller, *Int. J. Man-Mach. Stud.*, 7:1-13 (1974).

20. M. Braae and D.A. Rutherford, Theoretical and Linguistic Aspects of the Fuzzy Logic Controller, *Automatica*, 15:553-577 (1979).

21. W.J.M. Kickert and H.R. Van Nauta Lemke, Application of a Fuzzy Controller in a Warm Water Plant, *Automatica*, 12:301-308 (1976).

22. P.J. King and E.H. Mamdani, The application of Fuzzy Control Systems to Industrial Processes, *Automatica*, 13:235-242 (1977).

23. R. King and F. Karonis, Expert Systems in Industry, *AMSE Review*, 2:59-64 (1985).

24. P.M. Larsen, Industrial Applications of Fuzzy Logic Control, *Int. J. Man-Mach. Stud.*, 12:3-10 (1980).

25. E.H. Mamdani, Application of Fuzzy Logic to Approximate Reasoning Using Linguistic Synthesis, *IEEE Trans. Comput.*, C-26, 12:1182-1191 (1977).

26. E.H. Mamdani, Advances in the Linguistic Synthesis of Fuzzy Controllers, *Int. J. Man-Mach. Stud.*, 8:669-678 (1976).

27. N.J. Mandic, E.M. Scharf and E.H. Mamdani, Practical Application of a Heuristic Fuzzy Rule-Based Controller to the Dynamic Control of a Robot Arm, *IEE Proc.*, 132(D):190-203 (1985).

28. S.G. Tzafestas and N. Papanikolopoulos, Intelligent PID Control Based on Fuzzy Logic, *Proc. IFAC Symp. on Distributed Intelligent. Systems: Methods and Applications*, Varna (June, 1988).

29. I.G. Umbers and P.J. King, An Analysis of Human Decision Making in Cement Kiln Control and the Implications for Automation, *Int. J. Man-Mach. Stud.*, 12:11-23 (1980).

30. J. Van Amerongen, H.R. Van Nauta Lemke and J.C.T. Van der Veen, An Autopilot for Ships Designed With Fuzzy Sets, *Proc. 5th IFAC/IFIP Intl.. Conf. on "Digital Computer Applications to Process Control"* (1977).

31. W. Van de Velde, Naive Causal Reasoning for Diagnosis, *Proc. 5th Intl. Workshop on Expert Systems and Applications*, 1:455-473, Avignon, France (May, 1985).

32. S.G. Tzafestas, Knowledge Engineering Approach to System Modelling, Diagnosis, Supervision and Control, *Proc. IFAC/IMACS Symp. on SCS*, Vienna, 17-30 (Sept., 1986).

33. S.G. Tzafestas, A Look at the Knowledge-Based Approach to System Fault Diagnosis and Supervisory Control, in *System Fault Diagnosis, Reliability and Related Knowledge-Based Approaches* (S. Tzafestas, et. al. Ed.) 2:3-15 (1987).

34. S.G. Tzafestas, Artificial Techniques in Control, in *Proc. IMACS Symp. AI, Expert Systems and Languages*, Barcelona, 55-67 (June, 1987).

35. J.R. James, D.K. Frederick and J.H. Taylor, The Use of Expert Systems Programming Techniques for the Design of Lead-Lag Compensators, *Proc. Control'85*, Cambridge, England, 1-6 (1985).

36. T.L. Trankle, P. Sheu and U.H. Rabin, Expert System Architecture for Control System Design, *Proc. ACC*, Paper TP5, 1163-1169 (1986).

37. G.K.H. Pang, J.M. Boyle and A.G.J. MacFarlane, An Expert System for Computer-Aided Linear Multivariable Control System Design, *Proc. IEEE Symp. on CACSD*, 1-6 (Sept., 1986).

38. K.K. Gidwani, The Role of AI in Process Control, *Proc. ACC*, 881-884 (1986).

39. R. Sauers and R. Walsh, On the Requirements of Future Expert Systems, *Proc. 8th IJCAI*, Karlsrhue, W. Germany, 110-115 (1983).

40. C.G. Knickerbocker, R. Moore, L. Hawkinson and M.E. Levin, The PICON Expert System for Process Control, *Proc. 5th Intl. Workshop on Expert Systems and Applications*, Avignon, France, 1:59-96 (1985),

41. W.F. Kaemmerer and P.D. Christopherson, Using Process Models with Expert Systems to Aid Process Control Operators, *Proc. ACC*, 892-897 (1986).

42. L.A. Zadeh, Fuzzy Sets, *Information and Control*, 8:338-353 (1965).

43. L.A. Zadeh, Fuzzy Algorithms, *Information and Control*, 17:326-339 (1968).

44. L.A. Zadeh, Outline of a New Approach to the Analysis of Complex Systems and Decision Process, *IEEE Trans. Syst. Man and Cybern.*, 3:28-44 (1973).

45. L.A. Zadeh, Commonsense Knowledge Representation Based on Fuzzy Logic, *Computer*, 61-65 (Oct. 1983).

46. T.J. Procyk and E.H. Mamdani, A Linguistic Self-Organizing Process Controller, *Automatica*, 15:15-30 (1979).

47. S.G. Tzafestas, Fuzzy Expert Control: A Survey of Recent Results and Applications, ARWKBRC: *Advanced Research Workshop on Knowledge Based Robot Control*, Bonas (October, 1988).

48. P.J. Macvicar-Whelan, Fuzzy sets for Man-Machine Interaction, *Intl. J. Man-Mach. Studies*, 8:687-697 (1976).

49. R.L. Moore, L.B. Hawkinson, C.G. Knickerbocker and L. Churchman, A Real-Time Expert System for Process Control, *Proc. 1st Conf. on AI Appl.*, Denver, 569-576 (1984).

50. R. Sanz and A. Ollero, A Rule Based Inference Method for Supervision of Self-tuning Controllers by Using Microcomputers, *Proc. IFAC Symp. Low Cost Automation*, Valencia, Spain (1986).

51. A. Ollero and R. Sanz, Intelligent Workstation for Control and Supervision of Industrial Processes: Software Aspects, *Proc. IEEE Conf. VLSI and Computers* (W.E. Proebster and H. Reiner, Editors), 847-850, Hamburg (May, 1987).

52. S. Tzafestas, Applied Digital Control, *North-Holland*, Amsterdam (1985).

53. T. Umeda and K. Niida, Process Control System Synthesis by an Expert System, *Control-Theory and Advanced Technology*, 2(3):385-398 (1986).

54. B. Porter, A.H. Jones and C.B. McKeown, Real-Time Expert Tuners for PI Controllers, *IEEE Proc. (Pt. D)*, 134(4):260-263 (1987).

55. N.R. Sripada, D.G. Fisher, A.J. Morris, AI Application for Process Regulation and Servo Control, *IEE Proc. (Pt. D)*, 134(4):251-259 (1987).

56. S. Tzafestas, M. Singh and G. Schmidt (eds.), *System Fault Diagnostics, Reliability and Related Knowledge-Based Approaches*, (Vols. 1,2), D. Reidel, Dordrecht (1987).

57. M. Singh, K. Hindi, G. Schmidt and S. Tzafestas (eds.), *Fault Detection and Reliabiltiy: Knowledge Based and Other Approaches*, Pergamon Press, Oxford (1987).

58. R. Patton, P. Frank and R. Clark (eds.), *Fault Diagnosis in Dynamic Systems: Theory and Applications*, Prentice Hall Intl. (UK) Ltd. (1988).

59. A. Ichikawa and S. Kobayashi (eds.), Special Issue on Expert Systems and Fuzzy Control, *Control: Theory and Advanced Tech.*, 2(3):329-344 (1986).

60. H.R. van Nauta Lemke and W. De-zhao, Fuzzy PID Supervisor, *Proc. 24 IEEE Conf. on Decision and Control*, 602-608, Fort Lauderdale, Florida (December, 1985).

12

CAUSAL REASONING IN A REAL TIME EXPERT SYSTEM

Andy Paterson and Paul Sachs

PA Computers and Telecommunications
Rochester House, 33 Greycoat Street
London SW1P 2QF

1. INTRODUCTION

ESCORT (Expert System for Complex Operations in Real Time) is an expert system that deals with the problem of cognitive overload experienced by operators of process plant. The system analyses the plant data to identify control and instrumentation failures and provides the operator with advice on crisis handling and avoidance.

ESCORT was originally developed as a demonstration system to show the feasibility of developing a real time expert system. The demonstration system has been described at length elsewhere (Sachs et al., 1986; Sargeant, 1985).

Since the completion of the ESCORT Demonstration System (EDS) in February 1985 we have been investigating ways of enhancing the system to provide more flexibility in the problem solving approaches used and, in particular, to deal with non-local events and reasoning with time. This work was first described in (Paterson et al., 1985).

This paper brings up to date the work presented in the papers mentioned above.

2. THE PROCESS CONTROL DOMAIN

Processes in oil refineries, chemical plants, power stations etc. are nowadays largely controlled by computers or digitally based systems. These computer systems serve the following functions:

— to control the process

— to shut down the plant if it gets out of control

— to provide information to the operators to allow them to manage the plant.

This article first appeared in issue 4/87 of Journal A.

If a process variable (eg. level, pressure, temperature) moves outside set limits despite the control system's attempts to control it then an alarm is sounded and the operator must take some action to remedy the situation. If he doesn't then some or all of the plant may automatically shut down.

2.1. Cognitive overload

In the past the values of process variables were displayed to the operator on standard size (eg. six inch by three inch) meters. The amount of data available to the operator was therefore limited by the number of meters that could be fitted on the panel in the control room. As a result only the more important variables were displayed in the control room. A computerised system, however, has no such limitations. Anything measured can be displayed to the operators. This results in a vast increase in the amount of data that the operators must cope with. For example, British Gas's Morecambe Field development will have the capability to handle over 30,000 items of data. The rate at which the process data changes can also be very great (500 lights went on and off in the first minute of the Three Mile Island incident (Baur, 1983). Trying to make decisions on the basis of large amounts of data that can change rapidly leads to cognitive overload with the possible consequence that the operator misperceives the real state of the process plant and hence takes incorrect "corrective" action, often causing some or all of the plant to shut down.

3. ESCORT

ESCORT is designed to ease the cognitive load on operators by applying expert knowledge to incoming data and providing advice. ESCORT's development was influenced by several design objectives. The system should:

— diagnose the underlying problem that caused an alarm

— indicate the relative priorities of all current problems

— group together all alarms caused by a single problem

— provide advice to the operator within one second of being informed of an alarm

— use only the data available on the process control computer's databus.

In developing ESCORT there were two basic technical problems to be overcome:

— how to deal with large amounts of constantly changing data in real time

— how to reason about the behaviour of a process plant.

The solution of these problems involved the development of a special architecture for a real time expert system and the use of causal reasoning. These two aspects of the system are discussed in depth in the following sections.

The ESCORT demonstration system was implemented on a Xerox 1108 running Interlisp-D and LOOPS. In order to show the system working a simulation of a process plant was written. This runs on a PDP11 in Pascal. Subsequent development work has been done using KEE (Knowledge Engineering Environment) on a Xerox 1186. The original simulation has been used to test out this work.

3.1. The process plant used by the ESCORT

The ESCORT demonstration system operates using data from a simulation of part of an existing North Sea oil platform. Part of the plant is shown in figure 1. The diagram shows two of the five vessels simulated, their connecting pipes, control valves and indicators. For clarity much of the instrumentation detail has been omitted (only that relevant to the examples presented in this paper is included). The purpose of the plant is to separate out natural gas liquids (NGL) from the mixture of gas, NGL and glycol input through the pipe on the left. The NGL is output from the bottom of VX02. The pipes on the right lead to heating and cooling systems.

The plant control system is configured as a number of control loops in which the position of the output valve is controlled by the process variable which it affects. For example, in VX01 there is an NGL level sensor (normally called a level transmitter) (LT0101). Its output goes to a controller which compares the level with what it ought to be (its set point). If the level is too high then the controller sends a signal to the valve (LV0101) to open further. If the level is too low then the valve is closed a little. There are also alarms (eg. LAII0101) that activate if the process variable becomes too high or too low. The combination of transmitter, controller, valve and alarms is known as a control loop.

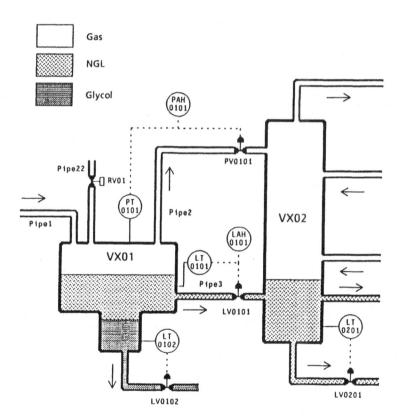

Fig. 1

— respond to new alarms quickly

— keep existing diagnoses up to date

— respond quickly to operator commands.

5. CAUSAL REASONING IN THE MAIN DIAGNOSTIC SYSTEM

The main diagnostic system's function is to arrive at and keep up to date diagnoses of plant problems. In order to produce a diagnosis of the problem that caused an alarm, ESCORT must be able to reason about the behaviour of the process plant. ESCORT currently uses a reasoning approach based on explicit knowledge of the cause and effect relationships that exist in a plant. This gives ESCORT two important capabilities that would be difficult to achieve using only heuristic "rules of thumb" for diagnosing problems. These are:

— dealing with non-local events

— reasoning with time.

5.1. Dealing with non-local events

A fault in a particular control loop can cause alarms to occur in both that control loop and other control loops (in which there may be no fault). The first of these we have termed local events–where the event (eg. an alarm) and the fault that caused it are in the same control loop. The second we have termed non-local events – where the event and the fault are in different loops. Non-local events are a major cause of cognitive overload since they can be far more numerous than local events and obscure the location of the real problem.

5.2. Reasoning with time

When trying to find the cause of an alarm it is often necessary to have a knowledge of previous events and problems in the plant. Put another way, a "snapshot" of all current data values may not contain sufficient information to form a correct diagnosis. This is especially true in the case of non-local alarms where, by the time the alarm occurs, the original problem may have been fixed or have completely recovered.

It is the ability to refer to past data and diagnoses in order to reason about the dynamics of the plant and understand what is currently happening in the plant that we have referred to as reasoning with time.

5.3. An illustration

This section contains a simplified example of a situation in which non-local events occur and reasoning with time is necessary. Some points should be noted about the following example. Firstly, the original problem is an operator error and all control systems function correctly. Secondly the rate at which gas is input to VX01 is constant. Figure 2 shows in a graphical form what happens to the pressures and levels involved (the numbers on the graph refer to the stages given below).

The sequence of events is as follows:

1. The operator accidentally opens pressure relief value RV01 for about 10 seconds.

2. The pressure in VX01 drops from 36 bars to 22 bars causing a low pressure alarm (PAL0101) to occur in VX01. The low pressure alarm

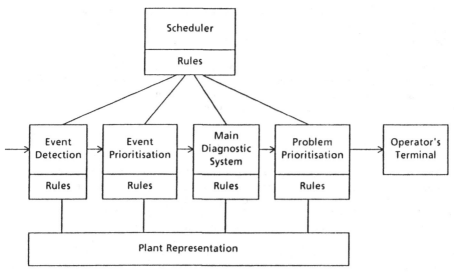

Fig. 2

4.3. The main diagnostic system

ESCORT's main diagnostic system has two basic tasks:

— to diagnose the underlying problem that caused a new event

— to keep current problems up to date.

This is the largest and most complex module in ESCORT. The operation of the main diagnostic system is based on causal reasoning and is described in depth in section 5.

4.4. The problem prioritiser

Once a problem has been diagnosed it is then prioritised relative to all other current problems. This prioritisation takes into account such factors as how urgently this problem requires operator action and the consequences of not rectifying the problem. Unlike the event prioritiser, the problem prioritiser has information about the underlying cause of the problem as well as its symptoms. This allows the prioritisation to be an accurate reflection of the actual alarm priorities. The output of this module is a prioritised list of problem diagnoses that is ready for presentation to the process operator.

4.5. The scheduler

Any system which has to handle large amounts of data and operate in real time must be able to schedule its resources effectively. To enable ESCORT to do this a knowledge based scheduler is used to determine which activity to perform next. The scheduler's rules take into acount:

— what new events have occurred

— the state of current problem diagnosis

— commands from the operator.

By using this approach of dividing the processing into different modules that can be explicitly scheduled ESCORT can:

Most of the problems in process plants are caused either by operator errors (eg. adjusting a controller set point incorrectly) or by failure of the control and instrumentation systems (eg. a control valve sticking). Failure of part of the plant itself (eg. a pipe rupturing) is considerably rarer. The faults that can occur in the simulation include control valves sticking open or closed, transmitters failing high or low, and level and pressure switches failing. In addition, the person operating the simulation may make errors in controlling the process.

4. A REAL-TIME EXPERT SYSTEM ARCHITECTURE

The majority of expert systems are knowledge based in the sense that they use explicit representations of domain knowledge. However, they often use an algorithmically defined reasoning strategy. Time critical systems require more sophisticated ways of determining reasoning strategy since they have to cope with widely varying computational loads, deal with multiple events in parallel, and respond quickly to new events. ESCORT therefore uses a knowledge based scheduler to provide the flexibility to allocate resources depending on the current state of its problem solving and the plant.

There are seven basic tasks that ESCORT needs to be able to perform:

1. recognise events in the process plant which may indicate a problem

2. prioritiese these events

3. analyse an event to infer the underlying problem

4. monitor existing problems to detect any changes in their status

5. prioritise the underlying problems

6. present problem diagnoses to the operator

7. respond to operator requests (eg. for explanation of reasoning).

Of these, tasks 1 to 5 use a knowledge based approach.

To reflect these different tasks ESCORT was configured as shown in figure 2.

4.1. The event detector

The event detector is the first expert system module on the route from data to advice. Its task is to recognise any plant data states which may indicate a problem and which should therefore be further investigated. Such data states are known as events.

Typical events refer to alarm states in the process plant. More complex events can refer to expectations of alarms (based on simple extrapolation of process variables). The output of the event detector is a list of events.

4.2. The event prioritiser

Before the list of events is passed on to the main diagnostic system the list is prioritised. At this stage the prioritisation can only be rather crude as the underlying problem in unknown. Priorities are therefore assigned on the basis of the relative importance of different areas of the process plant and the alarms within them.

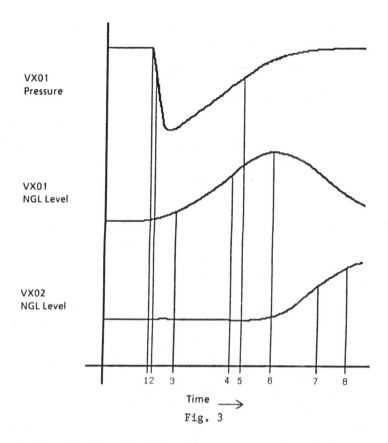

VX01
Pressure

VX01
NGL Level

VX02
NGL Level

12 3   4 5   6   7   8

Time ⟶

Fig. 3

is considered to be a local event since ESCORT associates relief valves such as RV01 with the loop controlling the same pressure.

3. NGL is normally driven from VX01 into VX02 by the pressure difference between the two vessels. However, the pressure in VX01 is now lower than that in VX02. The flow therefore ceases (there are one-way valves to stop reverse flow) causing the NGL level in VX01 to rise rapidly.

4. An NGL high level alarm (LAHH0101) occurs in VX01. This is the first non-local event.

5. The pressure in VX01 is recovering towards its set point (it has been rising gradually since RV01 was closed). The initial alarm (PAL0101) now ceases.

6. There is now sufficient pressure in VX01 to force NGL into VX02. As a result the NGL level in VX01 starts falling. The NGL level in VX02 starts rising. LV0201 opens 100% but cannot cope with the increased input from VX01.

7. A high NGL level alarm (LAH0201) occurs in VX02. This is the second non-local event.

8. As NGL empties out of VX01 the NGL level in VX02 continous to rise, eventually causing the process to shut down completely.

To summarise, one fault causes three alarms, one local and two non-local. The second non-local event causes the process to shut down. But by the time this second non-local event occurs the original cause and local event have disappeared. ESCORT's task is to relate all these alarms to the one cause. If only the snapshot of stage 8 is a available then there is no information available to indicate what caused the problem.

5.4. Overview of the causal reasoning approach

There are three basic components in the approach that has been used. These are:

— the plant representation: a definition of the process plant

— causal couplings: specifications of the cause and effect relationships in the plant

— a hypothesis network: ESCORT's current view of what is going on in the plant.

5.5. The plant representation

In order to reason about a process plant a representation of it is required. This can be thought of as comprising two parts: a definition of the components used to make up the plant and a representation of how these components are connected to each other.

Each type of component of the process plant is described by a class in a class inheritance lattice. Figure 4 shows a small excert of the class inheritance lattice used in ESCORT. The two main features that are relevant here are that a child class is considered to be a specialisation of the parent and that the child inherits the definition of the parent but can add to or change it. For example, a control valve is a specialisation of (or kind of) valve and therefore inherits the definition of valve which includes variables such as position (open or closed).

The actual items that make up the plant are represented by instances of the classes. Figure 5 shows, as an example, the level transmitter unit LT0101. It is the variables contained in a unit (termed "slots" in KEE) that specify the different types of relationships between the components (input, measurement, part of, etc.). For example, a pipe will have a slot named InputTo which will point to the vessel for which the pipe is an input. These slots together hold all the connectivity information for the plant.

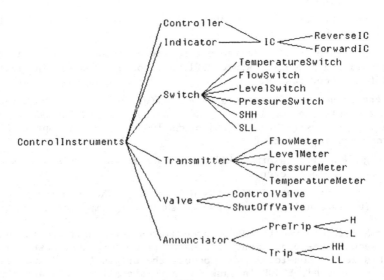

Fig. 4

The LT0101 level transmitter unit

Slot name	Value	Comment
PartOf	L0101	which control loop
Measures	NGL:01	what does it measure
InstrumentOutput	2.274	numeric value of transmitter reading
RateOfChange	-0.06	rate of change of transmitter reading
Reading	High	interpretation of transmitter reading
Max	2.5	maximum legitimate transmitter reading
Min	1.5	minimum legitimate transmitter reading
FaultStatus	OK	is transmitter working OK

Fig. 5

5.6. Causal couplings

The couplings specify the cause and effect relationships that occur in the plant. All couplings used in ESCORT are unified. A unified coupling is one that applies to all members of a class (eg. it is applicable to all transmitters not just a specific one). The use of unified couplings dramatically reduces the number of couplings that need to be defined.

ESCORT does allow plant-specific (ie. non-unified) couplings to be used if required, to define some relationship peculiar to the plant being dealt with.

An example of a unified coupling is shown below in figure 6. What this coupling means is:

If a level transmitter is reading high then one possible cause is that the liquid that it is measuring has a high level.

The links field of the coupling specifies the links that must exist between the cause and effect for this coupling to apply. For example in figure 6 the link constrains the cause to be the liquid that the transmitter measures. If the link definition were omitted the coupling would mean:

If a level transmitter is reading high then one possible cause is that any liquid in the plant has a high level.

5.7. The hypothesis network

The system's current view of what is going on in the plant is represented by a hypothesis network. Each hypothesis represents some assertion that may be true, false or unknown.

The hypothesis network is built up by applying incoming data to the causal couplings and plant representation. Hypotheses are created as a result of an event occuring in the plant (eg. an alarm). Once a hypothesis is created it is then tested to see if it is true. If the hypothesis is

Effect:	Reading of LevelTransmitter is High
Cause:	Level of VesselLiquid is High
Links:	LevelTransmitter measures VesselLiquid

Fig. 6

not testable its truth value is initially set to unknown; it may subsequently be reset by truth propagation. If the hypothesis is unknown or true then ESCORT uses the causal couplings to work out possible causes and then creates a hypothesis for each. These are then investigated in a similar manner.

An example of a hypothesis network is shown in figure 7.

Each box is a hypothesis. Within the box, the top line denotes the plant unit that the hypothesis refers to. The second line specifies the hypothesis as a variable/value pair. The third line specifies the truth value of the hypothesis and how that value was obtained: T = by testing, P = by propagation. Any hypothesis that has been greyed over is no longer true. At one time it was true; the time at which it ceased to be so is recorded on the bottom line.

The hypothesis network shown corresponds to step 4 of the example discussed in section 5.3, figure 3. It can be interpreted as saying the following:

Low pressure alarm PAL0101 was caused by low pressure in VX01 (now recovering) which was caused by relief valve RV01 being open (until 15:41). Low pressure in VX01 also caused low pressure difference between VX01 and VX02 causing low NGL flow from VX01. This caused a high NGL level alarm in VX01.

Note the way in which ESCORT has related two separate alarms back to the same underlying problem thus reducing the volume of information with which the operator has to deal.

5.8. Truth propagation

Not all hypotheses about the plant can be tested directly. For example, in figure 7 the reading of the pressure transmitter can be tested but the pressure itself cannot. Although we cannot test a hypothesis directly we still want to be able to come to some conclusions about it. This is in order to:

— construct a line of reasoning

— stop the generation of irrelevant hypotheses

— provide explanations to the user.

ESCORT uses a truth propagation scheme that is based on three assumptions about causal reasoning in process plants.

Assumption 1: If an effect is true then at least one cause must be

This assumes that the list of possible causes for a hypotheses is complete. In normal operation this should be the case. However, in the case of severely abnormal operation outside ESCORT's scope (eg. pipes rupturing) this assumption becomes invalid. ESCORT may therefore reach erroneous conclusions in such circumstances.

The practical value of this assumption is that if all but one cause has been shown to be false the remaining one can be assumed to be true. An example of this in figure 7 is that once the hypothesis PT0101-FaultStatus-Failow is shown to be false VX01-Pressure-Low can be assumed to be true.

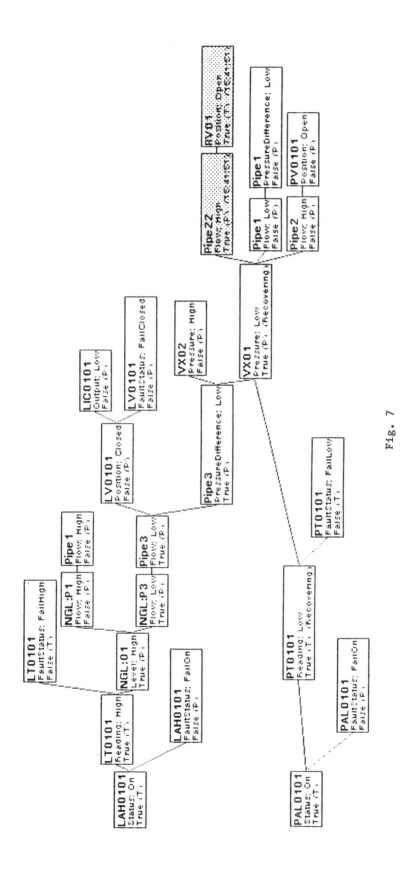

Fig. 7

227

Assumption 2: If a cause is true then so is its effect

An example of this in figure 7 is that once RV01-Position-Closed is true then Pipe22-Flow-Low also becomes true.

Assumption 3: There is only one cause for a hypothesis

This assumption is not true in all cases. However, the justifications for making it are:

— two simultaneous problems causing one alarm is a rare situation

— the major cause of cognitive overload is one problem and many alarms, not vice versa

— operators make this assumption; once they have fixed the first cause, only if the problem does not go away do they start looking for a second cause. This is what ESCORT will do.

An example of this is that when Pipe22-Flow-Low has been shown to be true (by assuption 2) the hypotheses regarding Pipe1 and Pipe2 are assumed to be false.

6. CONCLUDING REMARKS

The ESCORT demonstration system showed the feasibility of an expert system operating in real time. It also demonstrated the usefulness of an expert system to ease the cognitive overload experienced by process plant operators, even though its conclusions were limited to local faults.

The approach presented here has given ESCORT the ability to:

— maintain hypotheses about the past and use them when reasoning

— reason about how one control loop interacts with another

— correctly diagnose and explain the causes for non-local events.

All of these are achieved with the system operating in real time. As a result ESCORT's effectiveness in dealing with cognitive overload is greatly increased.

REFERENCES

1. P. Baur, "Strategies for eliminating human error in the control room", *Power* (May, 1983).

2. A. M. Paterson, P. A. Sachs and M. H. M. Turner, "ESCORT - The application of causal knowledge to real time process control", *Expert Systems 85*, (ed. M. Merry). Cambridge: Cambridge University Press, pp. 79-88, 1985.

3. G. Pitcher, "Expert systems mean less operator load", *Control Systems* (May, 1985).

4. P. A. Sachs, A. M. Paterson and M. H. M. Turner "ESCORT - An Expert System for Complex Operations in Real Time", *Expert Systems* (January, 1986).

5. R. A. E. Sargeant, "Experience of building a real time expert system", *Proc. Apollo Forum "An Introduction to Artificial Intelligence"* (1985).

13

KNOWLEDGE-BASED TOOLS FOR MANUFACTURING CONTROL: A GENERAL VIEW

József Váncza and Péter Bertők

Computer and Automation Institute
Hungarian Academy of Sciences
Budapest, Hungary

1. INTRODUCTION

The development of manufacturing technology and its integration into systems require a new approach. The control of individual machines under the supervision of human operators was a rather easy task, when the intelligence of men was used to cope with the problems arising in planning and manufacturing. Higher level of automation and the proliferation of flexible manufacturing systems have necessitated the introduction of more sophisticated control equipment and methods. The control hardware comprises of newer and newer computers, capable of executing complicated tasks, but the control methods and tools (especially the software) have a considerable delay in following this progress. Here an attempt is made to examine a new method, namely the application of knowledge engineering tools in manufacturing control. A comprehensive description of the subject is not intended here; instead, several specific topics have been chosen for discussion in some detail.

AI-related methods and ideas, as well as their possible applications are described. Existing state-of-the-art systems are also discussed, though we do not intend to introduce them with their full capabilities. Rather, underlying concepts and initiatives,that seem either to validate authentically the novel approaches considered, or to provide impulses for further research,are emphasized.

2. CONTROL TASKS IN THE MANUFACTURING ENVIRONMENT

The manufacturing process begins with the tasks of production preparation. The tasks of geometrical and technological design are not addressed here. Our starting point is the scheduling of operations. Then, after a short description of "operation control and monitoring", the topics of failure handling (diagnostics), and failure recovery are discussed.

2.1. Scheduling

The first problem is to decide about the events of the manufacturing process, i.e. to determine the routes of workpieces and the sequence of operations in the system. The scheduling activity assigns machines and other resources to workpieces, and allots the available "machine-time" to the jobs to be done. In standard systems this is a fairly simple combinatorial optimization problem, but the required flexibility in products and system operation significantly increases the complexity of the task; the variety of workpieces and lot sizes (and sometimes the mixture of lots) make the traditional methods ineffective[1,2]. The main advantage of a flexible system over traditional ones is the ability to manufacture different workpieces simultaneously, and adapt rapidly to different kinds and lot sizes of products. A flexible manufacturing system consists of several *stations* (machine tools or other equipment, e.g. measuring machines, workpiece buffers, etc.) and a *transportation system* connecting them together. The stations may consist of single-purpose, not convertible machines, such as simple lathes, or of multi-purpose machines, such as machining centres capable of holding large tool sets. The latter type of equipment provides greater flexibility at the price of making the system operation (scheduling included) more complicated.

Scheduling has to be performed at several levels. Usually two of them are used; *medium term scheduling* covering the throughput time of several dozen lots, and *short term* (or in some cases *real time*) *scheduling*. The medium term scheduling is based on the orders put to the factory, and its output is the demand put to the workshop. This means that the operation plan of the factory determines the starting time and due time of lots. The short term scheduling gives then the real sequence of events in the manufacturing system, and it may be invoked several times, if unexpected events (machine breakdowns etc.) occur, and a new timetable has to be set up for the operation. The medium term scheduling is a larger task and may be done off-line; since the time requirements are not so strict. The short term scheduling, however, is a much more time-sensitive problem; the rescheduling time may be critical because it may hinder the resumption of operation.

The scheduling has to operate within the framework of the factory, and there are several constraints which have to be observed. One group of constraints consists of resource and capacity constraints; the other group concerns production. One category of resource and capacity constraints concerns the fixed factors of the system, which cannot be modified or replaced. The number of machines and the available buffer size restrict the number of workpieces which can be present at a time in the system, either under machining or waiting for machines. In addition, there are modifiable constraints, e.g. several lots may need the same type of fixtures or tools, and a temporary increase in their number (by buying or borrowing them) may help in critical situations (provided they are available within the required time).

Production constraints are usually unmodifiable, because they are influenced by factors which are outside the authority of the scheduler. The *release time* and *due time* of lots may not be modified by the scheduler automatically. The only possible action is to indicate if a lot cannot be finished when it should be, but this can be regarded as inevitable, and these constraints are not inherently modifiable.

2.2. Production control

The normal operation of a manufacturing system is the step by step execution of the timetable generated by the scheduler. The normal operation poses no difficult problems; traditional programming methods can easily cope with it. The detection of breakdowns and other unexpected events during operation, however, is a more difficult matter, and a variety of methods have been developed to detect different kinds of events. Monitoring acts as a real-time feedback to the operation control, i.e. if an error is detected, other control functions are activated for failure diagnostics and recovery.

2.2.1. Monitoring

In a broader sense, monitoring is responsible for the maintenance of the dynamic data records used for event notification and decision information[3]. It watches and keeps track of the various - both unexpected and expected-events. Here, however, only the problem of failure detection will be examined. The basic purpose of monitoring is to detect a fault as soon as possible, and stop the operation of affected parts. There may be many kinds of failures, and different methods have been developed to detect different failures; they usually need sensors and signal processing devices. Depending on the system, monitoring is performed comprehensively during operation, or only some parts of the system are checked regularly. Continuous monitoring can be based on preliminary information about the process, or may detect abrupt changes in the monitored signal without any information about the process occuring on the machine[4]. In the latter case additional analysis of the sudden change is required, e.g. whether the cutting torque increase was due to a change in depth of cut or to a tool break.

2.2.2. Failure diagnostics

After detecting a failure the failed part must be localized. The damage must be assessed; the actual state of the system must be recognized and analyzed. Diagnostics is system dependent, and in most cases a human operator equipped with certain diagnostic aids can determine the failed part and find the way to fix it. In other words, the intelligence of a human being is used to solve the problem. No special skill is needed to fix most of the problems. An ordinary operator can usually perform well and only very few breakdowns need qualified service personel for the repair. Notwithstanding, the skill needed is very complicated for computers, mainly because of the shortage of information available in digital form.

2.2.3. Recovery

After localizing the cause of error, the next step is to resume the normal operation. Fully automatic recovery is a very difficult task, and in manufacturing it is not aimed at. The system may give advice to the operator how to treat the situation and how to fix, eliminate or circumvent the problem at hand. Then an automatic rescheduling of operation can adjust the system to the new situation.

3. GENERATING SCHEDULES

3.1. Least Commitment Strategies

The production should be able to handle both incessant external changes (e.g. rush orders) and possible internal instability of the shop floor. In principle, the best strategy is to generate production schedules on the basis of the current shop status information. Empirical studies have shown that the shop performance could be significantly improved by frequent rescheduling[5]. Based on the *least-commitment* principle, real-time schedulers react to changes promptly by updating schedules: they schedule within a limited time-frame, not beyond the validity of the available data warrants.

Two typical implementations of real-time schedulers advocating the
least commitment principle are the *"opportunistic"*(6) and the
"multi-pass"(7) schedulers. They both apply a composition of simulation
and rule-based reasoning methods. Short-term schedules are derived using
the current status information via rule-based inference. Additionally, in
order to provide the schedulers a short-term look-ahead ability, simulation is used to predict the future course of the controlled systems.

3.1.1. The Role of Simulation

In both cases, simulation of the manufacturing systems to be
controlled is based on models that are sufficiently fine-grained to
represent all the entities taking part in the "manufacturing game";
machines, transportation devices, buffers, tooling, fixtures etc. The
model is updated whenever the specifications of the real system change.

In an opportunistic scheduler the real and the simulation execution
of a previously compiled command sequence is performed concurrently.
Since simulation is faster, in the remaining time interval the next
schedule is generated by a rule-based scheduler departing from the state
predicted by the simulation pass. Thus, simulation is used to forecast a
starting point for generating the next schedule in the near future.

On the other hand, in the multi-pass scheduler, simulation helps to
choose from among several hypothetical scheduling alternatives suggested
by a rule-based scheduler. The virtual execusion of all the alternative
schedules enables a prospective analysis of future scenarions. So defi-
ciencies can be detected in the command sequences prior to their execution
and the scheduling alternatives can be evaluated and ranked by a suitable
evaluation function.

3.1.2. Rule-Based Scheduling

The paradigm of rule-based reasoning provides probably the most natural framework to represent and utilize dynamic dispatching rules and
scheduling heuristics that are almost exclusively applied whenever scheduling decisions should be taken without delay. However, it is not only
their form that suggests the direct use of rule-based methodologies.
Dispatching rules and heuristics are the results of empirical studies.
Consequently, they form an open, contradictory and incomplete set.
Opinions on the applicability of specific rules are quite different, too.
In practice, dispatching rules are set-up (or excluded) and tailored by
fine-tuning of their preconditions in the course of comprehensive applica-
tion-dependent experiments. The development of rule-based systems is
quite similar, they can be built up by the gradual incorporation and refinement of distinct pieces of knowledge.

A feasible way to organize knowledge encapsulated in the form of
rule is the loose decomposition of the whole rule-set i.e. the arrangement
of particular rules into contexts. Rules related to the same context may
interact with each other more often than with rules associated with other
contexts. Furthermore, strategic (meta-) rules can be introduced to select the current context and/or to resolve conflicts, when several (equally plausible) rules apply to a particular situation.

The rule-based parts of both kinds of real-time schedulers are organized according to this pattern. Dispatching rules, such as:
"a part having the shortest imminent process time is given the
highest priority",
can be described in a domain-specific rule-language(6) or in the clausal

form of first order predicate logic[7]. Contexts defined by the type of machine service being requested, and the origin of parts for servicing the request are used in[6]. Principles which reduce the complexity of the problem are also given in the form of meta-rules[7]. Additionally, criteria for selecting scheduling rules that conform best to a given situation and/or to a particular managerial objective are represented at the level of meta-rules.

Since real-time schedulers have to act in reply mode, at least a number of their rules should be used in a pattern-directed, forward-chaining manner. As a matter of fact, the opportunistic scheduler works solely in this way. The multi-pass controller is able to reason also backwards when it calls for further data. In these cases it establishes a goal that will provide the data and attempts to achieve this goal following the usual goal reduction scheme.

The *modus operandi* of the multi-pass controller exemplifies a more general idea about the combination of AI and deterministic simulation methods. This approach (discussed e.g. in[8]) suggests the use of AI techniques in an optimistic manner, allowing them to explore a broad terrain of hypothetical alternatives and taking no heed of contradictions, global optimum objectives. Simulation and related analytical tools, on the other hand, can be applied to resolve conflicts and evaluate the alternatives presented in a pessimistic, rigorous fashion.

Apparently, none of the real-time scheduling strategies can address the other key issue of scheduling, namely, the complexity of real-world problems. The complexity of scheduling problems, along the time dimension, may be greatly restricted by the least-commitment principle that precludes the possibility of long-term optimization. Actually, their look-ahead power cannot be arbitrarily increased, since it depends largely on exogenous factors such as the error rates of the distinct system components.

3.2. The Reformulative Method

Traditionally, a scheduling problem is solved so as to fit into one of the available mathematical schemes, and the solution is generated by optimization techniques (for a comprehensive survey of these algorithmic methods see[9]). In the case of a real-size problem, the correspondence between reality and model is, however, quite doubtful, and thus the optimum of a schedule is also questionable.

The reformulative approach to scheduling problems attempts to find satisfactory solutions of realistic problem settings. At any time, when it appears that a satisfactory solution cannot be reached, the model is restated and iteratively refined, until satisfactory solutions can be found relatively easily[10]. The more realistic and detailed the shape of a problem instance is, the more constrained and delimited is the space of its possible solutions. Making the model richer, by adding more and more real-life constraints, will make the scheduling problem less sensitive to particular solution methods[10,11].

The reformulative approach faces the problem of combinatorial complexity directly. Its most important representative is the ISIS system for job-shop scheduling[1], which paved the way for other knowledge-based schedulers[12,13].

3.2.1. Frame-Based Knowledge Representation for Realistic Problem Setting

The reformulative approach needs the support of a rich and flexible scheme for representing, if possible, every aspect of manufacturing knowledge related to scheduling. This knowledge can only be captured and exploited in symbolic representation and a well-structured hierarchical fashion. The scheme and language for its representation must be rich in conceptual primitives[14], whose role is to define the inheritance semantics for any relevant relation. The set of these primitives has to be open and expandable. They should support the accessibility and legibility of the encoded knowledge.

Today, the most efficient way to define a rich set of conceptual primitives that enables the step-by-step articulation of knowledge structures is the frame-based representation[14]. In advanced frame-systems taxonomic relations with inheritance can be defined freely. Additional inference tools, such as production rules are also available.

The modeling system of ISIS provides elementary concepts of states, objects and actions, related to each other by a rich choice of temporal and causal relations. By using these primitives as building blocks, one can describe the physical structure of a manufacturing system (including the static and dynamic features of the system and its components), the various manufacturing processes taking place, and the organizational structures and relations.

The main contribution of ISIS and its variations [12,13] to the progress of "scheduling" is, however, the methodology for capturing and handling technological and organizational constraints, altering production objectives. These are formulated within the general framework of constraints. Categories are defined for physical constraints, causal restrictions (e.g. operation alternatives), resource availability constraints, preference constraints, and organizational goals (e.g. due dates). For each constraint instance, it is determined how the degree of its satisfaction can be registered, how the constraint can be relaxed, what importance it has, and under which conditions it should be applied. A network of constraints is constructed by declaring interdependency relations, and establishing a mechanism for creating and propagating constraints.

3.2.2. Problem Reformulation via Constraint Directed Search

Schedules in ISIS are generated in an incremental fashion. Initially, a particular problem is posed in the form of a set of allowable routings for the orders to be processed. These routings specify operation precedence constraints, alternative manufacturing processes, and the substitutability of resources. This problem formulation is iteratively restated. Constraints are bound either by propagation or, if they cannot be satisfied, by relaxation during the course of a multi-level search process. At each level, a pre-search analysis selects those operations that span the space of all possible states representing alternative, partially unconstrained schedules. Then beam search is conducted, expanding a limited number of the good states. Alternative interim solutions are subject to a post-search analysis. If the solutions offered are innefficient, the search space is re-stated by the relaxation of one or more constraints.

Constraints play the key role in search for satisfactory schedules. They define the partial search spaces, and (like human schedulers) are the main medium for information storage and exhange. Also the stratification of the search process is determined, in essence, by the hierarchy of constraints. Though this is not a necessity, conflicts among constraints may

236

be highlighted by bottleneck analysis. Then those constraints which are most strictly bound, e.g. resources producing a bottleneck, can direct the search in an opportunistic way[12].

4. DETECTING FAILURES: MONITORING

At the stage of manufacturing control, where the execution of scheduled operations is supervised, the task of monitoring is to recognize whether the real performance of the system deviates from its planned - and expected - behaviour. This routine inspection task can be executed by using data- and function-based strategies. As a matter of fact, the co-existence and co-operation of several monitoring processes, having different underlying strategies, is a key of the fast and correct re-assessment of the status of a system.

4.1. Data-Driven Monitoring

The heart of this problem is to establish a correspondence between a particular data pattern - a mix of sensor and status signals, and projected values - and one or more modes of failure. It has to be decided whether one or more of the a *priori* defined symptoms of possible malfunctions can be recognized. The quality of data available to identify symptoms (the data might be noisy, unreliable and incomplete) makes the handling of uncertainties inevitable. The bottom-up approach implies that a large amount of data should be processed in real-time. Thus, fast, rule-of-thumb methods should be applied in order to guarantee the dynamism of the monitoring process.

Though a bit apart from manufacturing, experiences of recent AI-based applications to real-time process control[15,16] suggest how to handle the problem of *data-driven monitoring*. Knowledge used to detect particular malfunction symptoms should be organized as a loosely-coupled collection of chunks, each related to a particular error-mode. Commonly, such a knowledge chunk can be represented by means of production rules that span a cause-effect network for a subset of observations and a particular hypothesis. Nodes in the network represent considerations with belief (certainty) values, which are used to confirm or reject hypotheses, or to calculate the evidential weights of other nodes. The inference subnetworks for each symptom should constitute - by having connection points at the level of observations only - a wide, flat overall network. Belief is propagated along the hierarchy, so that the supporting/disconfirming evidence for a node is calculated from the belief values of the next lower level nodes in the hierarchy. Finally, the hypotheses which have gained strong enough evidence, can be picked up as conjectures of possible failure modes.

Driving the assessment of alarm hypotheses by such a simple control mechanism enables a fast evaluation of an observed data set against many hypotheses. It is easy to identify several failure modes at the same time. It might help considerably if, during the diagnosis phase, multiple causes of effects are discriminated.

However, this rule-of-thumb strategy is applicable only if a more systematic, deeper investigation process is executed concurrently. The set of malfunction symptoms cannot include all possible states that should be anticipated, and these descriptions themselves could be - as they generally are - imperfect, contradictory, too strict or even to permissive. As a matter of fact, one must not extend largely the circle of situations anticipated in this way, because it deteriorates the most fascinating capability of this reasoning method, i.e. the ability to react to the most critical situations without delay.

4.2. Function-Oriented Monitoring

In the heart of any monitoring system there is a hidden, implicitly encoded goal-structure (conceptualized by the designer of the system), the primary goals of which are: to maintain security, to guarantee system and process survial, and to preserve the system and its components from fatal damage. Of lower priority is the requirement (goal) of accomodating the production to a changed environment. However, the goals of specific monitoring activities should be recorded explicitly, in order to be able to reason about them, i.e. to evaluate, compare goals, and switch (if need arises) to another monitoring mode. In real-time systems such a capability is of crucial interest[17].

This inherently top-down approach implies an explicit, hierachical structure of monitoring goals. The goal-system can be established on the basis of the functional properties of the monitored system. Taking this approach, a hierarchy of various system functions has to be defined. The performance, proper or degraded, of these functions can be scrutinized from time to time in the light of the new incoming observations. If the *a priori* selected key functions are performed in an acceptable way, then it is reasonable to assume that the overall system also works properly. Looking at the example of a manufacturing cell, high-level key functions can be defined so that they globally take care of the flows of materials, energy and information within the cell, between the outer world and the cell, and finally check the co-function of cell operations. At the next lower level, more elementary functions may check whether workpieces, tools, fixtures and part programs are dealt with correctly.

The monitoring mechanism can be robust and reliable only if it is sensible enough and, at the same time, has the capability to inspect certain aspects of the operation of the controlled system. Thus the *event-based* and *function-oriented* monitoring processes should run concurrently. Paths of information exchange have to be defined in order to allow the data-driven process to interrupt the top-down checking of key system functions, whenever the sign of a severe malfunction is detected by one of the *ad hoc*.techniques. The required quick transition between the different monitoring processes can be facilitated by blackboards developed principally for the continuous co-ordination of distinct knowledge-sources which operate in real-time (see e.g. [18]).

5. DIAGNOSIS: EXPLOITING DEEP AND EMPIRICAL KNOWLEDGE

When a failure mode has been detected (or guessed), its root cause is sought. Thus diagnosis is performed in order to determine the location, the type and the severity of possible malfunctions, whose external manifestations have been detected previously. The link backwards from the effect to the cause is, however, rarely evident (especially in the case of multiple effects and/or causes); the cumulative causal relations have to be explored through more stages by uncovering interim causal paths. The spectrum of methods applicable to the assessment of fault hypotheses is very wide. For manufacturing purposes, however, AI-based diagnositic techniques have been applied only occasionally (e.g. see[19]). Here, the two extremes, namely *diagnosis based on model* (first-principles) versus *diagnosis using empirical (shallow) knowledge*, will be presented and the potential of an intermediate approach for manufacturing control will be discussed.

5.1. Model-Based Diagnosis

Diagnosis can be supported by an appropriate structural and behavioural model of the system. Based on this model, expectations on the internal operation of the system or its components can be derived, which, in turn, can be compared with the observed behaviour of the real system. Symptoms, i.e. discrepancies between the expected and observed performance, indicate malfunctions in the real system, provided of course that the model is correct. Principally, "only" the accurate description of the system structure and behaviour is needed, but no explicit fault-model is necessary. For example, only the knowledge about how a component of the system should operate is needed, whereas the methods, how to check it, can be derived from its specifications and design. Design descriptions of a device can be directly used in a resolution-like inference process to confirm or reject hypotheses about discrepancies between the actual and intended operation of the system where the device belongs[20].

In advanced model-based diagnosis systems a clear distinction is drawn between *behaviour prediction,* i.e. the task of a predictive inference component, and diagnosis performed by a separate domain-independent diagnostic engine[20,21]. At least ideally, such diagnostic systems are applicable to any case at which a man-made system is being diagnosed, although (due to their absolute dependence on the model) their scope of practical employment is quite restricted. In fact, the capabilities of model-based diagnostic systems are demonstrated almost exclusively in digital electronics, where an appropriate model can be set up relatively easily.

However, there are already available a number of results of these theoretical researches. They have provided techniques for handling multiple faults and performing incremental diagnosis. Succesfull attempts have been made to reduce the search for faulty components (i.e. to keep the number of components under consideration small) by introducing abstraction levels[20] and incorporating facilities for handling *a priori* fault-probabilities of components[21,22].

5.2. Empirical Diagnosis

Heuristic rules can be used to reduce the search for failure causes, and find plausible symptom-cause connections, instead of constructing causal paths. These rules assign symptoms to faults; they are based on empirical associations, and only occasionally refer to the anatomy, i.e. to the internal operation of the system diagnosed. This strategy needs no causal model, but an explicit model of faults. However, no fault-model can be all-inclusive. Thus, there is a possibility that the diagnoser, who relies solely on shallow knowledge, "falls from the knowledge cliff". Hence, the use of empirical techniques is recommended only if the knowledge about the subject is badly structured, diverse and vague, i.e. when an explicit model cannot be established for guiding the search from the observations to the sources of failures. Regarding technical systems, this is rarely the case.

In empirical diagnostic systems approximate reasoning has a global key role, not only in handling uncertain (mostly sensory) input information. Since it is often impossible to perform evidential reasoning in a precise manner, supporting/disconfirming weights should be assigned to the (interim) conclusions drawn. Features supporting approximate reasoning can be incorporated in different mathematical bases. However, there is always a chance that the misuse of the underlying probabilistic theory leads to erroneous probabilistic calculations.

239

5.3. Integrated Approach

Taking an example, in a manufacturing environment one of the hardest diagnostic tasks is that of the ón-line diagnosis of machine tools. Though their complex structure can be given up to the last screw as a composite of several mechanical, hydraulic, electrical and elecronic components, important causal relations describing their behaviour cannot be captured accurately beyond a certain level of detail. Probably, owing to the inneficient knowledge about the cutting process itself, most of the relations among observations (about the quality of the workpiece, internal parameters of the functional subsystems, acoustic emission, etc.) and functional changes remain vague, or can be stated only as rules of thumb. Thus on-line diagnosis of machining systems and processes is often solved without the real understanding of the underlying cause-effect relations[23].

One more issue calls for an integrated approach, namely the time required to generate explanatory hypotheses. In case of emergency, model-based reasoning would be too slow; reflex-like heuristics are surely needed when symptoms indicate catastrophic failures.

The paradigm of the "cooperating diagnostic specialists"[24,25], set up for purposes of medical diagnosis, but applicable also in technical domains, permits the dualism of empirical and model-based strategies. Within this paradigm the complexity inherent in a manufacturing (and in any other technological) system can be coped with in the most natural way, i.e. by describing the system and the related diagnostic knowledge on several levels of abstraction. The omission of superfluous structural details and the suppression of certain aspects of the system's behaviour facilitates reasoning about its operation only in terms of knowledge relevant at a given conceptual level.

In this framework, diagnostic knowledge can be articulated and organized by employing the structural architecture of a system as a hierarchy of specialists. Each system component is associated with a specialist, who possesses both the behavioural knowledge needed to discriminate the component's normal and faulty operation, and the structural knowledge about the decomposition of the given system component into more primitive elements.

When a specialist module is invoked by an alarm message, it attempts to establish or reject the hypothesis that a particular system component is defective. By rejecting a hypothesis all the subordinate fault-suspects are automatically discarded. However, once a fault-hypothesis can be established, structural knowledge is used to evoke subordinate specialists. Thus the diagnostic process can proceed torwards investigating more specific failure hypotheses at levels of finer resolution. This scheme corresponds quite well to a mental model describing how human operators cope with complex diagnostic tasks[26].

The above "establish/refine" inference process is actually a kind of classification[24]. Particular fault hypotheses can be corroborated and/or refuted either by evidential reasoning steps or by deep-reasoning (e.g. by local simulation), although how it is carried out is invariant to classification. A plausible policy is to use heuristic rules for common cases and go back to local models when no rule is applicable[27].

This diagnostic method supports the detection of multiple failure sources too; messages sent by the monitor can stimulate concurrent diagnostic processes at several nodes of the specialist hierarchy and also the refinement process may ramify at any node.

6. RECOVERY

Having detected manufacturing problems, control actions are to be initiated either to terminate abnormal conditions or to compensate for their consequences, until stability is achieved again. However, due to the risks of recovery, duties can hardly be performed without human assistance. The task of recovery is to recommend treatment methods, rather than to intervene automatically in the operation of a (partially) defective manufacturing system.

In the simplest case there is a repertory of fast responses in the form of fault-correcting actions. For example, a rule-based production system performing operation control tasks may have a section for handling errors and exceptions, as suggested in (28). Remedial actions are appended to control rules in (29). This "recovery tail" describes how to restore the states of devices, if a failure has occured during the execution of an operation sequence which has been initiated by a specific rule.

However, pre-established procedures, i.e. actions linked to particular control conditions, are not adequate for a system to recover from every possible situation. Decisions of human operators, who attempt to determine how to act in a given situation, are motivated by a more or less overt and structured set of goals. Naturally, the goals are of different weights; maintaining the safety of the overall controlled system and its components is of higher priority than, for example, saving semi-finished parts for re-use. These goals, behind the failure recovery logic, should be captured and expressed in an explicit goal-hierarchy, no matter if our intend is to aid operators or to replace them. As a matter of fact, a well-established hierachy of various recovery goals enables the operators, in case of emergency, to keep under control the sequential breakdown of system functions and prevent the system from total collapsing.

Recovery is a goal-oriented task of action planning in a dynamic environment. It involves a reactive, time-constrained mode of decision-making, so that both a set of possible decision (action plan) alternatives and a library of plan primitives should be ready for use.

In (18) a framework for goal-oriented recovery is outlined which seems to be applicable to the world of manufacturing. When the goal structure used for recovery is explicitly maintained, plan fragments can be attached directly to the goals, each one describing how to achieve a particular goal. Alternative ways may be offered by several plan fragments affixed to a goal. Once recovery is informed that due to unexpected failure(s) some particular goals cannot be accomplished, the highest level one (among the violated goals) is considered, and a sequence of remedial actions is sought. If in the given situation no alternative is executable, the next higher level goal has to be pursued. The approach advocates multi-level reasoning about goals to be achieved and about failed attempts to accomplish desired goals. This capability is essential when the normal operation of a large system has to be resumed(30).

7. APPLICATION LIMITS OF KNOWLEDGE-BASED TOOLS IN MANUFACTURING

It can be taken for granted that in real-life applications, such as manufacturing control, the "avant-garde" computing techniques will spread only if there is a chance to combine them with long-standing engineering frameworks. Traditional techniques are not (and cannot be) fully outdated by the new artificial intelligence tools.

For example, in the case of scheduling, robust and optimum-seeking algorithms of operational research should surely be adopted within any knowledge-based scheduler. There are various initiatives for integration of knowledge-based and traditional tools, especially in scheduling. In [31] the use of mathematical optimization techniques is suggested for generating long-term schedules on the basis of aggregated data, while coordinated expert systems, having detailed knowledge about parts of the overall manufacturing system (such as production lines, immediate storage and transportation systems), would be charged with the task of minimizing the differences between the optimal and current schedule. According to other views, AI methods should help in putting real problems into mathematical shape, and then into the qualitative analysis framework [10,32]. The optimistic use of knowledge-based techniques is also envisaged for generating many satisfactory alternatives, while, at the same time, operational research methods would resolve existing conflicts and consider optimality objectives in a more rigorous fashion [8].

Operational research gives also other possibilities to the AI research; it has a subtle taxonomy of well-defined problem types. A rich choice of solution schemes is available, and the best-fitting one can be selected by considering the types of constraints, specifications of variables, and optimization objectives.

A proposed framework of generic tasks in knowledge-based reasoning [25] enables the *ad hoc* taxonomy of AI-based tools to be reconsidered . The new approach is task oriented, in contrast with the one presently applied, which is rather implementation-technology oriented. Generic tasks are defined at an abstraction level, where suggestions on how the task-specific knowledge should be structured and organized, and on what kind of control regimes should be used, are derived from the nature of the tasks. The generic tasks – such as hierarchical classification, hypothesis matching and assembly, state abstraction, object synthesis and knowledge-based information passing – are atomic, and serve as building-blocks in constructing more complex architectures that support composite tasks. At this level of abstraction, the role of knowledge-based reasoning within the problem domain can also be analyzed. So the treatment is not constrained by the characteristics of particular techniques that provide the means to implement the generic tasks.

At the moment, the task-oriented exploration of potential AI tools, together with the detailed engineering analysis of manufacturing needs, are the prerequisites of further progress. The problem of matching tools and requirements is bound to recur. Recently, attempts have also been made to characterize the features and capabilities of existing expert system techniques in the context of typical manufacturing problems [33]. In [32] a structured taxonomy of the manufacturing problem domain, and the correspondence between the most crucial problems and AI-based tools has been presented. These systematic researches prepare the way for the next, more realistic phase, at which a standardized tool-kit tailored to the needs of manufacturing will be assembled.

8. CONCLUSION

The role of knowledge-based tools in manufacturing control has been examined. The AI-based methods that can and must be used to support scheduling, monitoring, diagnostics and recovery tasks in a flexible manufacturing environment have been discussed.

For the problem of scheduling two knowledge-based strategies have been presented. The first one, which is applicable to real-time scheduling, can be implemented by using rule-based techniques. The other method attacks the combinatorial complexity inherent in scheduling problems, and thus it needs the support of advanced knowledge representation techniques. For the monitoring problem, the mixed use of data-driven and top-down goal-oriented techniques has been suggested. In the case of diagnostics the idea of the "community of specialists" has been adopted. It permits the integration of specific empirical diagnostic techniques with a model-based method. Regarding the task of recovery, the importance of a clear goal-structure has been emphasized, and an action-planning scheme has been outlined.

Finally, the need for the co-existence of traditional and knowledge-based techniques has been discussed. This co-existence is a prerequisite for the proliferation of new methods and the practical implementation of experimental results.

ACKNOWLEDGEMENTS

The authors wish to thank their colleagues, Andräs Märkus and Zsöfia Ruttkay, for the many valuable discussions held with them.

REFERENCES

1. M. S. Fox, S. F. Smith, ISIS - A Knowledge-Based System for Factory Scheduling, *International Journal of Expert Systems* 1 (1), 25-49 (1984).

2. G. Bruno, R. Conterno, M. Morisio, The Role of Rule Based Programming for Production Scheduling, Working Paper, Politecnico di Torino, Italy (1986), p. 18.

3. J. McCahill, Towards an Application Generator for Production Activity Control in a Computer Integrated Manufacturing Environment, in: *ESPRIT'86: Results and Achievements* (Directorate General XIII, eds.), 891-900, North-Holland (1987).

4. P. Bertŏk, A System for Monitoring the Machining Operation in Automatic Manufacturing Systems, Ph.D. Thesis, Computer and Automation Institute, Hungarian Academy of Sciences, (1984), p. 151.

5. A. Ballakur, H. J. Steudel, Integration of Job Shop Control Systems: A State-of-the-Art Review, *Journal of Manufacturing Systems* 3(1), 71-79 (1984).

6. P. A. Newman, K. G. Kempf, Opportunistic Scheduling for Robotic Machine Tending, in: *Proc. of the Second IEEE Conference on Artificial Intelligence Applications*, 168-175 (1985).

7. R. A. Wysk, S. Y. Wu, N. S. Yang, A Multi-Pass Expert Control System (MPECS) for Flexible Manufacturing Systems, Working Paper, The Pennsylvania State University (1986), p. 28.

8. Zs. Ruttkay, A. Märkus, J. Väncza, The Place of Knowledge Engineering in Factory Automation, Working Paper E/41/87, Computer and Automation Institute, Hungarian Academy of Sciences, (1987), p. 20.

9. A. Kusiak, Application of Operational Research Models and Techniques in Flexible Manufacturing Systems, *European Journal of Operational Research* 24(3), 336-345 (1986).

10. J. J. Kanet, H. H. Adelsberger, Expert Systems in Production Scheduling, *European Journal of Operational Research* 29, 51-59 (1987).

11. T. J. Grant, Lessons for O.R. from AI: A Scheduling Case Study, *J. Opl. Res. Soc.* 37(1), 41-57 (1986).

12. S. F. Smith, P. S. Ow, The Use of Multiple Problem Decompositions in Time Constrained Planning Tasks, in: *Proc. of the Ninth Int. Joint Conference on Artificial Intelligence*, Los Angeles, USA, 1013-1015 (1985).

13. B. Sauve, A. Collinot, An Expert System for Scheduling in a Flexible Manufacturing System, *Robotics and Computer-Integrated Manufacturing* 3(2), 229-233 (1987).

14. M. S. Fox, Knowledge Representation for Decision Support, in: *knowledge Representation for Decision Support Systems* (L. B. Mathlie, R. H. Sprague, eds.) 3-26, North-Holland (1985).

15. R. L. Moore, L. B. Hawkinson, C. G. Knickerbocker, L. M. Churchman, A Real-Time Expert System for Process Control, in: *Proc. of the First IEEE Conference on Artificial Intelligence Applications*, 569-576 (1984).

16. M. L. Wright, M. W. Green, G. Fiegl, P. F. Cross, An Expert System for Real-Time Control, *IEEE Software*, 16-24 (March 1986).

17. A. Sloman, Real Time Multiple-Motive Expert Systems, in: *Expert Systems 85*, (M. Merry, ed.), 213-224 (1985).

18. D. Sharma, B. Chandrasekaran, D. Miller, Dynamic Procedure Synthesis, Execution, and Failure Recovery, in: *Applications of Artificial Intelligence in Engineering Problems* (D. Sriram, R. Adey, eds.) 1055-1071, Springer (1986).

19. M. S. Fox, S. Lowenfeld, P. Kleinosky, Techniques for Sensor-Based Diagnosis, in: *Proc. of the Eighth Int. Joint Conference on Artificial Intelligence*, Karlsruhe, West Germany, 158-163 (1983).

20. M. R. Genesereth, The Use of Design Descriptions in Automated Diagnosis, *Artificial Intelligence* 24, 411-436 (1984).

21. J. de Kleer, B. C. Williams, Diagnosing Multiple Faults, *Artificial Intelligence* 32, 97-130 (1987).

22. F. Pipitone, The FIS Troubleshooting System, *IEEE Computer*, 68-76 (July 1986).

23. R. Kegg, On-Line Machine and Process Diagnosis, *Annals of the CIRP* 33(2), 469-473 (1984).

24. J. Sticklen, B. Chandrasekaran, J. R. Josephson, Control Issues in Classificatory Diagnosis, in: *Proc. of the Ninth Int. Joint Conference on Artificial Intelligence*, Los Angeles, USA, 300-306 (1985).

25. B. Chandrasekaran, Generic Tasks in Knowledge-Based Reasoning: High Level Building Blocks for Expert System Design, *IEEE Expert* Fall 23-30 (1986).

26. J. Rasmussen, The Role of Hierarchical Knowledge Representation in Decision Making and System Management, *IEEE Transactions on Systems, Man, and Cybernetics* SMC-15(2), 234-243 (1985).

27. P. K. Fink, J. C. Lusth, J. W. Duran, A General Expert System Design for Diagnostic Problem Solving, *IEEE Transactions on Pattern Analysis and Machine Intelligence* PAMI-7(5), 553-560 (1985).

28. P. J. O'Grady, H. Bao, K. H. Lee, Issues in Intelligent Cell Control for Flexible Manufacturing Systems, *Computers in Industry 9*, 25-36 (1987).

29. O. Z. Maimon, Real-Time Operational Control of Flexible Manufacturing Systems, *Journal of Manufacturing Systems 6*(2), 125-136 (1987).

30. J. Rasmussen, L. P. Goodstein, Decision Support in Supervisory Control, in: *Analysis, Design and Evaluation of Man-Machine Systems* (G. Mancini, G. Johannsen, L. Martensson, eds.), 79-90, Pergamon Press (1985).

31. A. Kusiak, Artificial Intelligence and Operational Research in Flexible Manufacturing Systems, *INFOR 25*(1), 2-12 (1987).

32. A. Márkus, J. Hatvany, Matching AI Tools to Engineering Requirements, *Annals of the CIRP 36*(1), 311-315 (1987).

33. R. J. Mayer, P. G. Friel, M. Krishnamurthi, An Assessment of AI/ES Development Languages, Tools, and Environments for Manufacturing Applications, *SME Manufacturing Technology Review 2*, 213-239 (1987).

14

FAULT DIAGNOSIS OF CHEMICAL PROCESSES

Mark A. Kramer and F. Eric Finch

Department of Chemical Engineering
Massachusetts Institute of Technology
Cambridge, MA 02139 USA

1. INTRODUCTION

Operation of chemical and related process plants presents a range of important problems such as quality control, safety, availability, and efficiency. The computer, while well established in regulatory control functions, has had little impact on the broader range of process operations problems which are often difficult to formulate using conventional numerical/algorithmic approaches to computer programming. Knowledge-based systems offer an alternative to algorithmic programming and have been regarded as a promising approach to application of the computer in this domain.

One the first problems to be examined intensively in this context is fault diagnosis, the determination of the root causes of disturbances caused by equipment malfunction. The types of malfunctions associated with chemical plants include chemical effects such as introduction of impurities, occurrence of side reactions, and catalyst deactivation; mechanical faults such as valves sticking, blockages in pipes, poor mixing, fouling of heat transfer surfaces, and leaks; and information processing faults such as loss of signal, miscalibration of sensors, and controller failure. Timely discovery and identification of malfunctions is essential for the informed selection of the optimal response (viz. emergency shutdown, change in operating level, or issuance of maintenance requests).

In comparision to electronic, computer, or mechanical systems, process plants offer several unique diagnostic challenges. Some of the issues that must be treated by a process plant diagnostic system are as follows:

Detection of abnormal behavior: The system must discriminate between malfunction-induced disturbances and tolerable "backgound" disturbances. Large-scale plants are generally continuous in nature and operate at steady state, however, small disturbances and sensor noise cause regular disruptions of ideal steady state behavior. These variabilities can be described statistically. Deterministic models, in addition to statistical descriptions, are required to define a baseline of "normal" behavior for intentional transients (such as changes in production level in continuous plants) and dynamics associated with batch and semi-batch systems.

Plant Topology: Many plants incorporate recycle loops and complex stream topologies in their designs for optimal utilization of energy and raw materials. These designs defeat simple heuristics based on "looking upstream" for the failure origin. Further pathways of disturbance propagation are added by process control architectures involving feedback and feedforward flows of information.

Process Dynamics: Faults may trigger complex dynamic behavior in a process. Controllers in particular complicate diagnosis by attepting to compensate for undesirable changes induced by the fault. Inverse (normal ->high->low), compensatory (normal->high->normal), and oscillatory responses are observed in addition to simple monotonic deviations away from the steady operating state. Complex dynamics complicate the use of high/low/ normal patterns as a basis for identifying faults.

Fault Trajectory: Malfunctions can be abrupt or gradual degradations of performance, and markedly different symptoms can be exhibited depending on the rapidity and extent of the fault.

Process Sensors: In most process plants, measurements are collected and logged in a central control room. It is assumed that a diagnosis system will interface directly to the central process control computer for real-time access to process data. Measurements usually represent only a small fraction of the process variables and parameters that could be used in diagnosis. Due to technical and economic constraints, certain critical variables may be unmeasured, measured indirectly, or measured only periodically. The accuracy of different types of measurements may vary.

Thus, diagnosis of process plants presents some interesting challenges.. In this paper, several approaches to this problem are examined. In particular, we are concerned with the expert system methodology and how it applies in this domain. We then focus on model-based reasoning strategies, which due to the special nature of the process plant diagnosis problem, are sometimes favored over pure experience or expert-based techniques. In this regard, two specific methodologies that have been investigated in our laboratory are compared. The first of these is based on causal search of a graphical representation of process variable interactions and is implemented in the system DIEX (1). The second technique is based on functional analysis of the plant, using a representation of the plant as a set of intentional systems (2). This comparision gives some insight into the use of structure and function models as foundations for model-based reasoning in diagnosis. Finally, some suggestions are made for improving these methods through utilization of AI programming paradigms.

2. BASIC DIAGNOSTIC APPROACHES

Selection of a diagnostic strategy depends on the types and extent of knowledge available on the system being examined. This knowledge can be broadly classified into two categories: behavioral knowledge describing the operation of the system and diagnostic knowledge of how to diagnose the system.

In the process plant, it can be assumed that significant *a priori* knowledge on the behavior of the plant is available in the form of mass, energy, and momentum balances, heat and mass transfer correlations, and chemical reaction rate expressions. Also available are mental models of process behavior carried by operators and designers, which might be considered special classes of qualitative or semi-quantitative models, and general models of basic plant components (e.g. pumps, valves, pipes, heat exchangers, etc.). Despite the relative abundance of behavioral knowledge, certain portions of a process may be only poorly characterized (often the case for process chemistry), and the remaining models are subject to varying degrees of inaccuracy. Detailed knowledge of process behavior under abnormal conditions such as may result from faults is less readily available. .

Diagnostic knowledge is generally not available at the design stage and must be acquired through experience or by converting behavioral knowledge into diagnostic knowledge. The quantity of diagnostic knowledge available is primarily a function of the experience and training of the operating staff, which varies from plant to plant. Diagnostic knowledge is more valuable than behavioral knowledge since it can be used immediately in solving diagnosis problems. Behavioral knowledge cannot be used directly in diagnosis but must be used in conjunction with a diagnostic methodology that in effect converts behavioral knowledge to diagnostic knowledge.

Based on these sources of knowledge, we can see two approaches to diagnosis. The first is the expert system paradigm, utilizing diagnostic knowledge. The second approach is model-based reasoning, which is based on behavioral knowledge of the process. The former approach provides a proven, straightforward methodology for encoding existing diagnostic knowledge. Among the advantageous features of this approach are:

— **Declarative representation**: Knowledge is represented explicitly and transparently and thus it is easily codified, verified and revised.

— **Explanation**: The expert system can justify its recommendations so that the operator does not have to accept the advice of the computer on the basis of faith alone.

— **Extensibility**: The scope of the system can be gradually increased concurrent with the emergence of new cases and problems.

— **Model independence**: Expert systems can be applied to problems where fundamental principles are lacking and only heuristic solutions are available.

The power of this paradigm is evidenced by many successful applications in a wide variety of fields (3,4). However, expert systems suffer a number of disadvantages that include:

— **Requirement of pre-existing solution**: The current generation of expert systems lack the capacity for learning and must be supplied solutions to problems. Because potential faults are numerous and occur infrequently, there is often an inadequate opportunity to build up sufficient knowledge for this approach.

— **Development cost**: The effort of building expert systems in this domain is quite substantial, and due to design differences, the effort of constructing a diagnostic system cannot usually be amortized over multiple plants.

— **Completeness and reliability**: Faults cannot be induced solely for testing purposes. This implies the need for systematic and disciplined approaches to diagnosis.

In view of these factors, we believe that conventional expert systems *cannot satisfy all industrial needs in this area*. The model-based reasoning paradigm may be better suited to the particularities of process engineering and operations problems.

Model-based reasoning involves the following principal features:

1) A method or methods of representing processes and their behavior,

2) A general diagnostic methodology consistent with the representation of process behavior,

3) General or prototypical models, transportable between plants.

Clearly a major benefit of this paradigm is the reduction of the reliance upon heuristic rules. Another advantage is that development cost of the system is reduced by the use of transportable components and by utilizing behavioral knowledge in place of difficult-to-acquire diagnostic knowledge. In the following, two methods of model-based reasoning for process plant diagnosis are examined.

3. DIEX

The Diagnostic Expert (DIEX) (1) is a prototype system written in Franz Lisp that employs models of the cause-and-effect interactions between process parameters and state variables to address the diagnosis problem. It is in the spirit of the digraph technique first described by Iri et al. (5) but differs in many details of implementation, particularly in the treatment of object programming to support generic models.

3.1. Digraph Models

The basic model used by DIEX is a causal directed graph (digraph). Nodes in the digraph represent process variables and parameters and assume the qualitative states high (+), normal (0), and low (-). The normal range is chosen to represent the expected range of a variable under normal (fault free) operating conditions, to an arbitrary degree of confidence.

Digraph arcs represent directional causal interactions between process variables. A change in the variable represented by the *initial node* is transmitted by a physical mechanism to the variable represented by the *terminal node*. Each arc has the attributes sign, magnitude, and time. The sign attribute represents the relative direction of change of the initial and terminal nodes, a "+" sign indicating deviation of initial and terminal nodes in the same direction, and a "-" sign indicating deviation in opposite directions. The time attribute models the dynamics of deviation propagation, a "0" indicating zero or negligible delay between deviation of the initial and terminal nodes, and "1" indicating a positive time delay between deviations. The magnitude attribute is used only to remove branches from the generic models which may be unimportant in a particular context, by assignment of a "0" magnitude.

The causal interactions modeled by the digraph arcs are *local* in the sense that they link variables that are adjacent at the given level of modeling detail. Global constraints such as overall mass and energy balances, and global dynamics such as the loop transfer functions of feedback control systems, cannot by modeled directly using the digraph model.

A causal digraph is not unique and must be tailored specifically to the diagnosis task. A digraph is suitable for fault diagnosis if the primary effect of each potential fault is the deviation of a single digraph node, and if the digraph represents the physical system on a level of detail that maximizes resolution between faults. Palowitch (1) outlines a methodology for producing digraphs suitable for diagnosis.

One disadvantage of the rule-based expert system approach cited earlier is the lack of generality which requires a new knowledge base for each application. Although a new digraph is required for each DIEX application, the object-oriented programming paradigm (flavors in Franz Lisp) allows generic digraph models of common plant units (tanks, pipes, controllers, etc.) to be instantiated and linked together to model dissimilar processes. Only process topology, certain branch attributes, and the normal operating range of measured variables need be specified for each new application.

3.2. Fault Diagnosis using Digraphs

The diagnostic task is to locate the *root node* in the digraph which is deviated through the direct action of the fault. DIEX uses the assumption that only a single fault is present in the process at any time to simplify the search procedure. Under this assumption, diagnosis involves searching the digraph for a single node, termed a *primary deviation*, which explains the signs of all measured abnormal nodes. Because only a small proportion of digraph nodes are measured, the diagnosis procedure usually finds multiple primary deviations, any one of which could be the root node.

DIEX first generates potential primary deviations by searching causally upstream of a measured deviated (abnormal) node for all nodes that could potentially cause the observed deviation. This backwards search is terminated if 1) the edge of the digraph is reached, 2) the search loops upon itself, or 3) the search encounters a measured normal, non-controlled node. If more than one deviation is observed, a backward search is conducted for each deviated node and the intersection is taken to produce the set of primary node candidates.

DIEX accepts abnormal measurements in the order in which they occur. As new deviated measurements are entered into the system, a backward search is conducted from the new deviation and the resulting node set is intersected with the existing set of primary node candidates to update the diagnosis. This algorithm necessitates the strictly sequential deviation of variables along a path of disturbance propagation, otherwise the search may be prematurely terminated.

After potential primary deviations are located through graphical search, DIEX tests these candidates using global constraints (such as an *a priori* specification of dominant pathways in cases of competing inputs to a node), time delay information, and certain heuristics related to control loop function. To complete the diagnosis, DIEX lists (in no particular order) potential faults associated with each primary node and relates this information to the operator for further action.

4. SYSTEM-LEVEL DIAGNOSIS (SLD)

An alternate approach to diagnosis, based on the composition, function , and dependencies of process systems, has recently been proposed by the authors (2). The SLD shares many of the concepts of the digraph methodology but differs in several respects, most notably, the level of abstraction at which the process is modeled. The method uses functionality as the basis for diagnosis, rather than causal topology.

4.1. Process Graph Models

At the center of the SLD is a representation of a process in terms
of intentional (goal directed) systems. Each system is responsible for a
single goal important for process operations such as supplying a utility
stream, carrying out a chemical transformation, or controlling a temperature. A system is composed of lower-order unit functions which contribute to the fulfillment of the higher-order functional objective. A unit
function is any one of several duties a unit may perform. Faults may des-
troy or reduce the ability of a unit to perform one or more of its functions, producing system failure. Often, a single unit function is associated with two or more systems, and failure of the function results in
failure of any or all the associated systems. How a specific unit failure
affects a particular system strongly depends on the nature of the failure
and the type of system involved. Some systems are resilient and can continue to meet their objectives in the presence of certain failures, while
others fail in the presence of any unit function failure.

Three types of system are identified: passive (open-loop), control
(closed-loop), and external. The objective of passive and control systems
is to maintain a controlled variable at a specified level. Control systems achieve their objective by influencing the process through regulation of a manipulated variable or variables. Passive systems, lacking a
regulation mechanism, achieve their objective by maintaining the functionality of their constituent components at a nominal level. External refers to any system, passive or control, on the periphery of the process,
which for diagnostic purposes is treated as a passive system.

System performance is assessed by comparing measured values of a
system's controlled variable with an appropriate reference value, and in
the case of control systems, by comparing measured manipulated variable
values with expected values computed by the diagnosis system. Reference
values for controlled variables are setpoints assigned by operators in
the case of control systems and design values in the case of passive systems. For manipulated variables, several varieties of expected values
can be computed depending on the type of information included in forming
the expectation. Possible information sources include previous plant
experience, setpoint and design values, measurements of other controlled
and manipulated variables, measurements of the process environment, and
known operator actions. The simplest type of expected values, and the one
used predominately in this paper, are the steady state values of the manipulated variables. These are the same reference values used by DIEX.

Passive systems assume two qualitative states: *functional* if the
system's controlled variable is not deviated from its reference value,
and *malfunctional* if the system's controlled variable is deviated. Control systems assume four qualitative states:

1) *Functional* if neither the controlled or manipulated variables are
   deviated.

2) *Stressed* if the manipulated variable is deviated but the controlled
   variable is not deviated.

3) *Uncontrolled* if the controlled variable is deviated but the manipulated variable is not deviated.

4) *Saturated* if both controlled and manipulated variables are deviated.

States (2), (3), and (4) are all considered malfunctional states.

Dependencies between systems are depicted in the *process graph* as directed arcs. A control system with a single controlled and manipulated variable can have two types of associated arc: a controlled variable (CV) arc and a manipulated variable (MV) arc. Dependencies are considered *active* if the associated manipulated or controlled variable is deviated and *inactive* if the variable is not deviated. Only active arcs support disturbance propagation.

In summary, systems have two sources of dependency – internal dependency on the unit functions comprising the system and external dependency on other systems, depicted in the process graph. All system malfunctions can be traced to failures in one of these sources.

4.2. Fault Diagnosis using the Process Graph

The SLD proceeds in two stages. First, the process graph is examined to determine which systems are *source systems*, or systems directly affected by the fault (analogous to root nodes in the digraph). Then, generic rules are applied to determine which system components may be responsible for the malfunction. Again, the single fault assumption is used to expedite diagnosis.

4.2.1. Source System Identification

Despite the similarities of root node and source system location, the latter is more complex since a single fault may produce multiple source systems. In instances where multiple source systems are suspected, the problem is often one of discriminating between *possible* source systems which may or may not be genuine source systems, and *definite* source systems. A definite source system is indicated if a malfunctional system has no active propagation path directed toward it. Possible source systems are indicated if there is a set of malfunctional systems forming a closed loop with no active inputs.

4.2.2. Detailed Diagnostic Rules

After identification of possible and definite source systems, detailed rules are applied to narrow the diagnosis to specific components of the source system. The ability to distinguish definite source systems from possible source systems is critical in providing the maximum degree of fault discrimination at this stage of diagnosis. If one or more definite source systems are identified, the results of the generic rules obtained for each definite source system can be intersected to obtain the final diagnosis; otherwise, the union of the results obtained for all possible source systems must be used.

The detailed diagnostic rules depend on division of system functions into four classes:

1) Control functions (CNT) of the controller and control valve.

2) Control loop sensing functions (CSEN) of the controlled variable sensor in a control system.

3) Sensing functions (SEN) of the sensors (including control loop sensing functions).

4) Process functions (PRC) which include all non-control, non-sensing functions.

Control systems contain all four classes of functions; passive systems contain only PRC and SEN functions.

The detailed diagnostic rules narrow the failure candidate set by identifying which of the four function classes, if malfunctional, could result in the observed state of the source system and its neighboring systems. The complete rule set is given in Finch and Kramer (2). Application of these rules is essential if certain "latent" faults are to be captured in the diagnosis. For example, a small bias in a controlled variable sensor may produce a disturbance in neighboring systems while leaving the true source system in an apparent functional state. This type of fault is accounted for in the diagnostic rules.

5. DIAGNOSIS EXAMPLE

DIEX and SLD represent dissimilar philosophies in diagnosis of faults. DIEX is based on structural analysis, specifically, on the causal connectedness between process variables. SLD is based on analysis of specific losses of functionality. It is therefore interesting to compare the performance of these two approaches, using results obtained from both methods solving identical diagnostic problems.

5.1. Application of Methodologies

In the original design of DIEX given in (1), measured deviations (alarms) are entered into DIEX in the order they occur. Only the initial sign of the deviation is used, regardless of whether the variable subsequently returns to normal or changes sign. In this respect, DIEX has a memory for both the sequence and sign of past alarms, and requires a "reset" after each fault event. Under ideal conditions, this approach can enhance DIEX's ability to distinguish faults by making full use of the method's heuristics and bounding mechanism. However, in some fairly common circumstances, memory is a handicap, producing inaccurate diagnoses. False alarms (transient alarms not associated with a fault) and out-of-order alarms (alarms that occur out of sequence along a path of propagation) are two instances where strict observance of alarm sequences can produce inaccurate diagnoses. In contrast, SLD uses a "snapshot" of the current states of the process systems each time the diagnosis is updated. This memoryless approach allows the method to forget past false and out-of-order alarms. To remedy the problems experienced by DIEX when confronted with out-of-order alarms, alarms can be entered as if they all occurred simultaneously. This scheme is referred to as the *modified* DIEX application.

5.2. Example System

The process used for comparison is a continuous stirred tank reactor (CSTR) system, shown in Fig. 1. A first order exothermic reaction takes place in the CSTR, cooled by recycle through an external heat exchanger (HTX). Temperature, level, and recycle flowrate are regulated by feedback control systems (cascade control in the case of the temperature loop). Fifteen variables are measured[1], including all controlled and manipulated variables as required by the SLD methodology. A numerical dynamic simulation of this system was developed to simulate process faults. For each type of fault, the extent versus time profile of the fault could be varied

---

1. SP-SENSOR (setpoint of the FRC-2) and R2-SENSOR and R3-SENSOR (the valve stem positions for control valves CV-2 and CV-3), are the remaining measurements not shown in Fig.1.

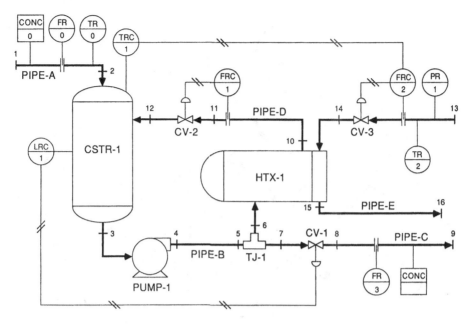

Figure 1 Process example.

to provide a realistic range of responses. Alarm bands were established
in relation to simulated sensor noise so that about one false alarm every
600 sensor-minutes was generated during normal operation (i.e. one false
alarm every 40 minutes).

The digraph model for this process contains 132 nodes and 184 arcs
(not shown). The process graph model, shown in Fig. 2, contains 10 nodes
and 16 arcs. Systems are defined in Table 1A, and system components are
described in Table 1B.

5.3. Analysis of Results

Although both diagnostic methodologies compared in the previous section use graph models to represent process behavior, differences in the
model representations lead to diagnoses with markedly different characteristics. Modeling at the structural level in DIEX produces a listing of
specific faults, while the SLD provides possibly faulty units and unit
functions. To compare the performance of the methods, a relationship must
be established between faults described at the system level and individual faults. This correspondence is given in Table 1C.

Relevant measures of performance are *accuracy* and *discrimination*
(relative sensitivity is not an issue since both methods use the same
alarm bands). Accuracy refers to the inclusion of the true fault in the
set of possibilities, and discrimination refers to the ability of the
method to exclude spurious faults from the candidate set. Ideally, a method   should be able to produce an accurate diagnosis with high discrimination.

Tables 2A–C list the results of three simulated fault runs. The size
of the fault candidate sets for DIEX and SLD after each new alarm   or

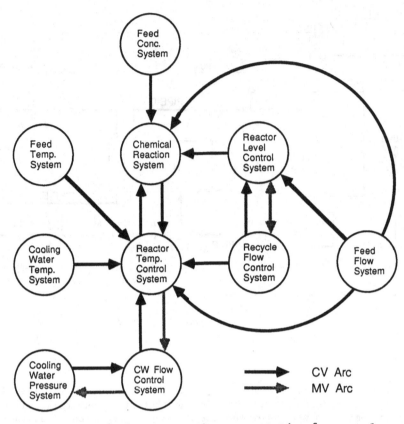

Figure 2   Process graph representation for example.

TABLE 1A. System Definitions

name	system description	system type
S1A	reactor temperature control system	control
S1B	cooling water flow control system	control
S2	reactor level control system	control
S3	recycle flow rate control system	control
S4	chemical reaction system	passive
S5A	cooling water temperature system	external
S5B	cooling water pressure system	external
S6A	reactor feed flow rate system	external
S6B	reactor feed temperature system	external
S6C	reactor feed concentration system	external

TABLE 1B. System Components

unit name	unit functions	func. type	system(s)
CSTR-1	fluid containment	PRC	S2
	fluid conductance	PRC	S2,S3,(S6A)
	fluid mixing	PRC	S4
	reaction (catalysis)	PRC	S4
	insulation	PRC	S1A
PUMP-1	pressurization	PRC	S2,S3
	fluid conduc./cont.	PRC	S2,S3
	insulation	PRC	S1A
HTX-1	tube-side containment	PRC	S1A,S1B,S2,S3,S4
	tube-side conductance	PRC	S1B
	shell-side containment	PRC	S2,S3,S4
	shell-side conductance	PRC	S2,S3
	heat transfer	PRC	S1A
CV-1	flow regulation	CNT	S2
	fluid conduc./cont.	PRC	S2,S3
	insulation	PRC	N/A
CV-2	flow regulation	CNT	S3
	fluid conduc./cont.	PRC	S2,S3
	insulation	PRC	S1A
CV-3	flow regulation	CNT	S1B
	fluid conduc./cont.	PRC	S1B
	insulation	PRC	S1A
TJ-1	fluid conduc./cont.	PRC	S2,S3
	insulation	PRC	S1A
PIPE-A	fluid conduc./cont.	PRC	S6A
	insulation	PRC	S6B
PIPE-B	fluid conduc./cont.	PRC	S2,S3
	insulation	PRC	S1A
PIPE-C	fluid conduc./cont.	PRC	S2
	insulation	PRC	N/A
PIPE-D	fluid conduc./cont.	PRC	S2,S3
	insulation	PRC	S1A
PIPE-E	fluid conduc./cont.	PRC	S1B
	insulation	PRC	N/A
FRC-1	computation	CNT	S3
FRC-2	computation	CNT	S1B
LRC-1	computation	CNT	S2
TRC-1	computation	CNT	S1A
CO-SENSOR	measurement	SEN	S6C
FO-SENSOR	measurement	SEN	S6A
TO-SENSOR	measurement	SEN	S6B
L-SENSOR	measurement	CSEN	S2
CA-SENSOR	measurement	SEN	S4
CB-SENSOR	measurement	SEN	N/A
T1-SENSOR	measurement	CSEN	S1A
T2-SENSOR	measurement	SEN	S5A
F1-SENSOR	measurement	CSEN	S3
F2-SENSOR	measurement	CSEN	S1B
F3-SENSOR	measurement	SEN	S2
P1-SENSOR	measurement	SEN	S5B
R2-SENSOR	measurement	SEN	S3
R3-SENSOR	measurement	SEN	S1B
SP-SENSOR	measurement	SEN	S1A

Note: conduc./cont. = conductance/containment

257

TABLE 1C: Relation between SLD functions and DIEX faults

unit type	failed functions	corresponding DIEX faults
CSTR	fluid containment	liquid leak
		vapor leak
		leak at outlet [1 outlet]
	fluid conductance	inlet blockage [2 inlets]
		outlet blockage [1 outlet]
	fluid mixing	none
	reaction (catalysis)	side reaction occurring
		catalyst fouling
	insulation	fire
		insulation removed
PUMP	pressurization	entrained vapor
		broken shaft or coupling
		power too high to motor
		loss of power to motor
	fluid conduc./cont.	leak
		blockage of suction or discharge
	insulation	fire
		insulation removed
HTX	tube-side containment	leak in cold stream
		leak between shell and tubes
	tube-side conductance	blockage in cold stream
	shell-side containment	leak in hot stream
		leak between shell and tubes
	shell-side conductance	blockage in hot stream
	heat transfer	severe fouling
		fire
CV	flow regulation	failed open
		failed closed
	fluid conduc./cont.	leak
		blockage
	insulation	fire
		insulation removed
TJ or PIPE	fluid conduc./cont.	leak
		blockage
	insulation	fire
		insulation removed
CONTROLLER	computation	failed high
		failed low
		setpoint high
		setpoint low
SENSOR	measurement	failed high
		failed low

TABLE 2A. Results for Reactor Temperature Sensor Bias (+2 $^{\circ}$C)

time (s)	alarm events	system events	DIEX std/mod	SLD	int[†]
			Number of faults		
1	T1-SENSOR High	S1A Uncontrolled	44/44	4*	0*/0*
5	SP-SENSOR High	S1A Saturated	43/43	28	9/9
11	F2-SENSOR High		18*/18*		0*/0*
18	R3-SENSOR Low		18*/34		0*/9
178	CB-SENSOR Low		18*/34		0*/9
236	CA-SENSOR High	S4 Malfunctional	18*/34	34	1*/10
323		S1A Stressed		14	1*/2
∞			18*/34	14	1*/2

* Actual fault not in set
† int = intersection of DIEX and SLD candidate sets

TABLE 2B. Results for Gradual Rise in Activation Energy

time (s)	alarm events	system events	DIEX std/mod	SLD	int[†]
			Number of faults		
136	CB-SENSOR Low		41/41		
157	SP-SENSOR Low	S1A Stressed	35/35	30*	2*/2*
168	CA-SENSOR High	S4 Malfunctional	21/21	14	3/3
180	T1-SENSOR Low	S1A Saturated	21/43	34	3/11
182	F2-SENSOR Low		9*/15*		1*/1*
209	R3-SENSOR High		9*/30		1*/11
472		S1A Stressed		14	1*/4
∞			9*/30	14	1*/4

* Actual fault not in set
† int = intersection of DIEX and SLD candidate sets

TABLE 2C. Results for Control Valve CV-2 Stuck (60% closed)

time (s)	alarm events	system events	DIEX std/mod	SLD	int[†]
			Number of faults		
1	F1-SENSOR Low		4/4	40	4/4
	R2-SENSOR High	S3 Saturated	4/4	40	4/4
22	T1-SENSOR High	S1A Uncontrolled	4/4	0*	0*/0*
31	SP-SENSOR High	S1A Saturated	4/4	40	4/4
48	F2-SENSOR High		4/4		4/4
50	R3-SENSOR Low		4/4		4/4
92	CA-SENSOR Low	S4 Malfunctional	4/4	40	4/4
101	CB-SENSOR High		4/4		4/4
108		S1B Uncontrolled		40	4/4
356		S1B Saturated		40	4/4
∞			4/4	40	4/4

* Actual fault not in set
† int = intersection of DIEX and SLD candidate sets

system state change are shown. DIEX results are shown for both the standard and modified applications. Inaccurate diagnoses are indicated by asterisks.

In runs 1 and 2, the F2-SENSOR alarm registers before the R3-SENSOR alarm, an example of an out-of-order alarm. Using the standard application of DIEX, this produces a series of inaccurate diagnoses persisting for the duration of the fault. Using the modified application, DIEX produces an inaccurate diagnosis only for one alarm period.

The SLD produces inaccurate diagnoses for short intervals during runs 1 and 3 because of transient uncontrolled states caused by the delay between disturbance propagation into a control system and controller action. Another temporary inaccuracy occurs in run 2, when systems S1A and S4 "ring" out-of-order.

In general, the SLD and the modified DIEX application performed equally well, producing mostly accurate diagnoses with approximately the same number of fault candidates[2]. A notable exception is fault 3 for which DIEX outperfomed SLD by a substantial margin by taking singular advantage of the process structure. However, the SLD produced the diagnoses with a far simpler system model than DIEX, indicating a significant benefit from the functional analysis approach.

6. TOWARDS AN IMPROVED ARCHITECTURE FOR PROCESS DIAGNOSIS

In view of the representational differences between SLD and DIEX, it is not surprising to observe variances of discrimination and accuracy. These differences are reflected not only in raw numbers of fault candidates, but also in the identities of the faults contained in the candidate sets. If both methods are accurate, then the actual fault will be in the intersection of the candidate sets produced by the two methods. Since in many cases the intersection is significantly smaller than either set individually (see table 2), discrimination of faults could be improved by a joint structural-functional approach that employs both model representations.

In addressing this problem, we are also confronted with a wide variety of industrial process types and designs. Inevitably, there are aspects of process behavior that are not captured by models. A diagnostic system cannot be completely effective without an ability to bring plant-specific knowledge to the diagnostic task, in the manner of the conventional expert system. The ideal diagnostic system should integrate structural, functional, and experiential knowledge from process experts.

The need for multiple knowledge sources has been recognized previously by several researchers. One of the authors has presented a method of integrating structural and experiential knowledge that involves converting structural information to pattern recognition rules that can be integrated with experiential rules using rule chaining and certainty factors (6). This approach has the disadvantage of having model information bound up in rules, and therefore not directly accessible to verification or

---

2. The absolute numbers of faults in Table 2 are more a function of the level of description than the quality of the diagnosis. For example, in a list of faults provided by DIEX, there may be four faults representing different modes of pump failure, five faults referring to blockage at different locations along a pipe run, etc.

modification. Additionally, the pattern recognition rules are not robust to variations in alarm order. Other workers include Davis (7), who presents an analysis of the complementary roles of structure and function modeling in electronic equipment diagnosis, and Fink and Lusth (8), who present an architecture for integration of functional and heuristic knowledge for mechanical and electronic systems diagnosis. The domains considered by these authors do not involve complex dynamics, and their techniques cannot be applied directly to process plants.

One of the major difficulties involved with diagnosis of process plants is out-of-order alarms. Alarm patterns depend on the amplitude and time trajectory of the fault, and in general there is no way to set alarm thresholds to guarantee strict ordering of alarms (1). In DIEX and SLD, serious problems in handling out-of-order alarms are encountered. In the previous section, a modified approach to DIEX was proposed to deal with the problem; however, inaccurate diagnoses could not be entirely eliminated.

Our present investigations are aimed at the problems of multiple knowledge sources and dynamics. We believe these issues can be addressed from the standpoint of knowledge representation and computer architecture, employing AI programming techniques. The key design concept is that of an *event interpreter*. An event is defined as any significant change or trend of a measured variable, or a significant change or trend in an aggregate quantity, such as a mass balance or a calculated efficiency. In the previous section, events of interest were crossings of high and low thresholds. This simple classification method is less than ideal. For example, a variable undergoing a discernable trend but within its alarm limits is considered normal, and a short transient alarm is treated the same as a strong persistent deviation. There is no conceptual difficulty in using techniques of statistical quality control to derive normal/abnormal classifications, or using more subjective descriptions such as intermittent, probable, strong, fast, slow, or oscillating to describe measurement trends.

The event interpreter operates by analyzing each event in order of occurrence. Knowledge appropriate to the type and identity of the event is applied, and an interpretation of the event is rendered. Diagnosis of the state of the plant is the sum of the interpretations of the individual events. An event interpreter achieves the integration of problem solving methods by viewing events from multiple perspectives. For example, an event can be interpreted from a topological viewpoint. A rule base containing general topological rules is brought to bear on each new event, and classifies the event as a direct consequence of transmission through a causal pathway from a previous event, a possible out-of-order event, or a primary deviation. It also determines if the event rules out or lends support to any previous fault hypotheses. Using this approach, robustness to variations in alarm order can be produced with a relatively simple rule base, as demonstrated by Corsberg (9).

An event can likewise be viewed from a functional perspective. Functional analysis determines if a change in a system's state is the result of other abnormal system states. The functional rule base also identifies from the four generic system functions (CSEN, CNT, PRC, and SEN) which could be responsible for the observed abnormalities. The functional analysis helps build support for hypotheses derived from topological considerations.

The plant-specific knowledge about faults and events available from operators and engineers can also be applied in an event-oriented fashion.

This knowledge is contained in rules that relate symptomatic events to particular faults. Classes of experiential rules are triggered by the appearance of key symptoms.

A version of the event interpreter has been implemented in a hybrid AI programming environment that combines rules, frames, and object-oriented programming[10]. Factual knowledge on the structure and function of the plant is represented in frames. These frames form the static data base that is addressed by event interpretation rules. Trend checking and event identification is implemented by *sensor monitors*. These monitors "watch" a particular measurement or alarm, and take action (triggering event interpretation rules, changing other slot values) when a significant change is observed. Monitors are implemented by demons associated with slots of the sensor objects. Demons (also called active values) are procedures that are executed whenever the slot value is changed. These monitors act autonomously and with the appearance of parallelism so that an independent monitor can be assigned to each sensor or alarm. Significantly, this distributed architecture leads to automatic treatment of simultaneous, non-connected events such as multiple non-overlapping faults, or the simultaneous occurrence of a fault and a false alarm.

7. CONCLUSION

This paper has summarized problems and solution techniques associated with fault diagnosis of chemical and process plants. The most important features of this problem are: complex topologies and dynamics, variability of symptoms based on fault trajectory, knowledge of process expressed as models, and incompleteness of diagnostic knowledge. These factors motivate an integrated approach utilizing structural, functional, and experiential knowledge. The basic methodologies for structural and functional analysis of process plants are contained in the DIEX and SLD systems, however, these approaches have several weaknesses that are demonstrated through an example. Recommendations are made for a new architecture utilizing a hybrid AI programming environment which address the current problems of DIEX and SLD.

8. ACKNOWLEDGEMENT

The authors gratefully acknowledge the research contribution of Dr. B.L.Palowitch Jr. This work was conducted with the support of the National Science Foundation under grant CBT-8605253.

9. REFERENCES

1. B.L.Palowitch Jr., Sc.D. Thesis, Massachusetts Institute of Technology, (1987).

2. F.E.Finch and M.A.Kramer, Narrowing Diagnostic Focus by Control System Decomposition, *AIChE J.*, <u>34</u>, 25, (1987).

3. P.Harmon, Inventory and Analysis of Existing Expert Systems, *Expert Systems Strategies*, *2*, No. 8, 1 (1986).

4. H.P.Newquist, True Stories: The Reality of AI Applications, *AI Expert*, *2*, No. 2, 63 (1987).

5. M.Iri, K.Aoki, E.O'Shima, and H.Matsuyama, An Algorithm of Diagnosis of System Failures in the Chemical Process, *Comp. Chem. Eng.*, 3, 489 (1979).

6. M.A. Kramer, Integration of Heuristic and Model-Based Inference in Chemical Process Fault Diagnosis, IFAC Workshop on Fault Detection and Safety in Chemical Plants, Kyoto, Japan (1986).

7. R.Davis, Diagnostic Reasoning Based on Structure and Behavior, in D.G. Bobrow (Ed.), *Qualitative Reasoning About Physical Systems*, Elsevier, Amsterdam (1984).

8. P.K.Fink and J.C.Lusth, Expert Systems and Diagnostic Expertise in the Mechanical and Electrical Domains, *IEEE Trans. Sys. Man. Cyber.*, SMC-17, 340 (1987).

9. D.Corsberg, Alarm filtering: Practical Control Room Upgrade using Expert Systems Concepts, *InTech.*, No. 4, 39 (1987).

10. M.A.Kramer, Automated Diagnosis of Malfunctions Based on Object-Oriented Programming, J. Loss Prevention Proc. Indus., 1, 247 (1988).

15

ARTIFICIAL INTELLIGENCE IN COMMUNICATIONS NETWORKS MONITORING,

DIAGNOSIS AND OPERATIONS

L. F. Pau

Technical University of Denmark
DK 2800 Lyngby, Denmark

1. INTRODUCTION

Network "reliability", more properly specified via its availability (percentage of time the network is available to its users) and its integrity (probability that data transmitted is not lost or damaged), is essential for all users. Moreover one should never forget that the "reliability" performance may, under some distributed applications, be more important than the general network capabilities.

Whereas this goal is generally achieved through a variety of techniques, all have in common the requirement for on-line data communications measurement, monitoring and diagnostic capabilities {21,24,23}. The classical techniques contributing to availability and/or integrity, are:

— distributed processing and distributed data bases with check words facilities

— redundant node processors (with switching between spare modules)

— redundant communication lines (with switching)

— message, datagram or packet routing strategies {14,19,22}

— communications prorocols {6,7,8,15}

— distributed flow control and resource allocation {5,9,10}

— high hardware reliability, software reliability, and on-line testability

— high reliability and integrity of the measurement, monitoring and diagnostic information transmission channels (including the lines for clock synchronisation)

— error detection in buffer and temporary storage units

— network supervisory control

— message generation {24} for tracing routing, traffic and condition information; essentially, a test/monitoring message or frame
is inserted every p such normal data message/frames to convey
information between nodes

— use of data messages as test stimuli, e.g. for the links {24}.

The above techniques represent state-of-the-art approaches to
communications networks monitoring, diagnosis and operations. It can be
seen that few among them incorporate procedural information, that is
information about the way in which the network operations are organized
under various conditions to achieve various goals, including network
reliability. Such procedural information, as well as search for
alternative goal fulfillment paths, are especially well handled by
artificial intelligence techniques. Essentially, knowledge representation
procedures, higher level inference and control structures, combined with
object oriented programming, and AI languages, allow for the incorporation
of procedural information and knowledge based search into the operations
or service functions of communication networks (switched networks, data
communications, radio communications, image distribution).

This chapter will be covering the following specific implementation
areas:

— failure detection and troubleshooting at the link and node level,
including failure detection features and their processing;

— intelligent contents-based message routing through the network;

— knowledge based message terminal distribution and archival;

— intelligent front-ends and help facilities to information
services;

— network and node system configuration;

— knowledge based security and access verification;

— analysis and monitoring of network status, incl. node
instrumentation.

Such implementation opportunities for artificial intelligence
techniques {33,34}, have been mostly investigated in relation with
research or actual fielding for:

— command and control information networks, with heterogeneous
users, nodes and transmission links {35,36,37,38,58}

— alarm and monitoring network for technical facilities (power
networks, avionics, data communications networks) {37,38,39,40,
58}

— office automation networks {41}

— electronic funds transfer networks {38,40}.

The major challenges are:

i. the measurement, monitoring and diagnostic capabilities implemented in distributed, imbedded or stand-alone test and monitoring units (abbreviation: TMU), which use the network itself or other links to communicate

ii. specific and distributed knowledge bases about the network, and goals to be achieved by AI search/reasoning, viewed as inferences on these knowledge bases

iii. implications on open/closed information system architectures.

2. FEATURES ACQUIRED BY THE TEST AND MONITORING UNITS

These features are at three levels (Fig. 1):

A: Analog signals
D: Digital signals
T: Traffic measurements

2.1. Analog signals

These analog features and the corresponding detectors are always important for noisy channels; they reach utmost ctiricallity in broadcast, microwave, satellite and fiber optics links:

A1: impulse noise counters, and time dependent distribution of this type of noise for some specified maximal threshold level;

A2: signal burst processors, which detect sequences of impulse noise;

A3: detection of fading and dropouts, and of their length distribution over time; a dropout is an interval of T or longer where peak-to-peak signal falls A (dB) or more below normal full carrier signal strength;

A4: signal distorsion level, especially of the signal carrier;

A5: overspeed variation;

A6: operating margin {31}, defined as the difference, in dB, between the current signal/noise ratio at repeater input, and the expected value for some specified bit error rate (BER);

A7: phase variations at the reception of some periodic signalling bit;

A8: margin bias obtained by amplifying crosstalk noise between parallel channels {29}.

2.2. Digitized signals

These features are easy to extract, and their detectors are of simple design, thus enhancing their widespread use in a node:

D1: bit error rate (BER); direct error counting is however slow, so that recursive approximations are often preferred {32};

D2: character error rate;

D3: block error rate;

D4: serial bit burst processors, detecting n-bit bursts in serial streams; e.g.: trains of 1's, excess 0's, impulse bits...;

D5: bit or symbol interferences {17} estimated on test messages, and relation to the BER;

D6: errors detected by the check-sum field and/or flag field in most packet frames (X - 25, etc...);

D7: memory, receiver and modem built-in test outputs;

D8: hardware condition codes received from the local and other TMU's;

D9: hardware and line condition codes received from other nodes and TMU's;

D10: number of circuit disruptions detected by the data link control procedure;

D11: excess zero's, one's densities {30}.

It is essential that these features be detectable regardless of the transmission mode (asynchronous, synchronous) and of the transmission protocols: TTY, BSC, SDLC, NRZI, HDLC, BDLC, DDCMP, ADCCP, X-25, SNA, TCP, D-format datagram (INWG - 6 - 1 of IFIP).

2.3. Traffic measurements

The node architecture must allow for traffic measurement instrumentation {16}, both for synchronous and asynchronous transmissions; delays are usually estimated via the time sub-field in the test messages or packets:

T1: periodically reinitialized counters measuring:
— negative message/packet acknowledgments from other nodes
— number of deadlocks {13}
— requests sent to other nodes (requests to send)
— tasks generated and processed locally in the node

T2: buffer utilization {11};

T3: empty buffer space;

T4: processing delay within each node;

T5: propagation/transmission delays between nodes;

T6: response time (incl. processing, propagation, routing and hand-shake);

T7: request to send delay;

T8: clear to send delay.

3. PREPROCESSING OF THE TEST AND MONITORING FEATURES

Each TMU must fulfill, in addition to measurements (A, D, T), two equally important functions, which are error compensation at the signal and data levels, followed by control and reconfiguration orders generated locally and eventually transmitted to other nodes. The bit or block error compensations should in principle be implementable anywhere in the node.

3.1. Error compensation {18} requires added node instrumentation:

a) At the signal level, this is done essentially by high capacity and high speed signal strorage, combined with improved signal detection and filtering, plus requests to repeat the broadcast.

One approach is to carry out adaptive source model estimation by a Viterbi-type algorithm, in order to characterize the noise on-line, and thus to assess BER. - A more simple remedy is to modify the signal detection threshold, it's offset, and to whiten noise.

b) At the bit level, a variety of error-detecting and correcting codes are universally used; the bit stream is generally first analyzed by a division by some finite length n - bit polynomial; the corrected signal is thereafter synthesized, and errors are detected by the generated check word:

— HDLC

— cyclic redundancy routines (CRC), e.g. $X^{16} + X^{12} + X^5 + 1$ for 16 bits words (ISO) and HDLC protocols

— multiple error-correcting binary codes for non-independent errors, such as fire codes (n = 32, 35, 48, 56)

— long block error correcting codes in conjuction with time-spreading interleaving logics

— high speed codes (convolutional codes with optimum distance) {27,28}

— multiple track correction.

Group errors are well detected by cyclic codes, but the frequency of undetected errors remains $10^{-7} - 10^{-10}$.

c) A very effective error control technique for bursty channels is forward error correction coupled with data interleaving before transmission and after reception; this causes the bursts of channel errors to be spread out and thus to be handled by the decoder as is they were random errors treated as specified in b).

d) At the modem level, bit forward error correctors may be used in case of RF links: BCH, Golay, Hamming, Interleaving.

3.2. Control and reconfiguration {4,24} lets the TMU software activate stand-by spare units locally, call back-up software, and start reinitializations.

Moreover, the TMU sends messages to the local flow controller, especially for priority reassignment, and buffer and link utilization (capacity, transmission rate).

Some of the orders or condition codes are also sent to neighboring nodes and the network supervisor (if any), for routing, recovery, and congestion control.

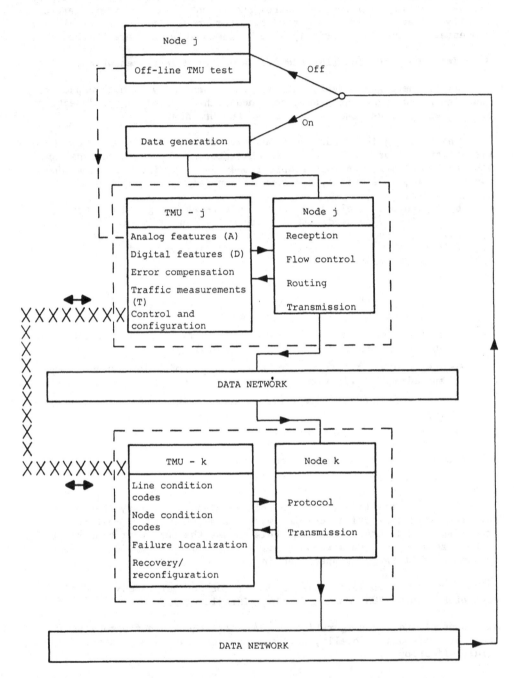

Fig. 1 Test and monitoring features.

4. COMMUNICATIONS NODE INSTRUMENTATION AND COMPROMISES

a) Although elementary microprogrammed error-detection-correction codes cover a large proportion of minor errors, those problems caused by high data flow networks with "irregular links" (microwave, satellite, fiber optics) are much more complex, and require extensive TMU and node instrumentation. To grasp the extent of these difficulties, let us mention that a satellite or fiber optic link may dump data at rates in excess of 120 Mbauds into a network, whereas distributed host computers connected to the network only accept, say, 25 Mbauds; any error leads to instantaneous network congestion and loss of data integrity.

However, even in low-to-medium capacity networks, there may be a similar requirement for extensive, although simpler, node instrumentation, motivated by relatively high BER of some units or peripherals. A BER=10^{-5} errors/bit (1 error every 20 s) for a modem at 4,8 kbauds is not untypical, and at least error compensation is required, if not TMU supervision.

In any event, error detection and correction will have to be implemented on the high density data recorders associated to the TMU (signal dropouts, errors) (e.g. G.C.R. codes) {25,26}.

At the supervisory level, very fast error detection is paramount to data integrity, and also to network or node reconfiguration orders {3} all of which cannot be generated by a human operator.

Moreover, to understand a sequence of events, it is necessary to obtain a trace/history of relevant events; this justifies the fact that the operator may access the TMU and its memory through triggered data dumps, selective feature recording, real-time access to trafic features, and video enhancements.

b) These requirements impose a number of compromises, of which the most important are:

   i: TMU hardware complexity (incl. storage capacity), and speed ∨∆ TMU reliability

   ii: TMU reliability {2} ∨∆ network integrity and availability

   iii: TMU generated traffic through the network ∨∆ actual data traffic and transmission delay (all overheads excluded)

   iv: Node and TMU testability ∨∆ complexity

   v: Restoration of node operations via back-up procedures in case of TMU failure.

c) The best approaches to improved network "reliability" seem to have been local on-line TMU monitors, coordinated via a network supervisor. The extent of this supervision varies however considerably from one existing network to another. The TMU architectures should not affect normal node operations, and should comply with the following design rules:

— high speed TMU's, operating much faster than each individual node

— large localized TMU storage capacity;

— knowledge based test and monitoring message generation (see Section 7);

— standard TMU design, regardless of transmission codes (EBCDIC, ASCII, BAUDOIT) and protocols.

5. FAILURE DETECTION PROBABILITIES AND DATA INTEGRITY {12}

a) In most existing prototype or experimental data networks, the introduction of TMU's and/or network supervisor has clearly improved the data integrity. If the BER of the worst node or TMU subsystem is B (errors/s), and if the transmission channels and equipment conditions are the same, networks with localized TMU measurements seem to offer BER at the network user level of 10^{-3} to $10^{-6} \times B$ (errors/s). Because of the multiplicity of factors involved, e.g. reduced deadlocks due to routing, it is however difficult to provide more precide evidence of data network measurement capabilities.

b) Another compromise involves bandwidth or data rate. An error compensation techniques which employs a rate 1/2 code (100% redundancy) will require the double bandwidth of an uncoded signal. On the other hand, if a rate 3/4 code is used, redundancy is 33% and bandwidth expanded only to 4/3. Error correction moreover reduces the S/N ratio for a given BER by 5-6 dB compared to systems without error compensation.

c) The BER vs S/N ratio compromise also depends on the modulation method used {1}, regardless of pre- and postprocessing:

— frequency-shift keying FSK (coherent/non coherent)

— phase shift keying PSK

— pulse-code modulation PCM (polar/unipolar)

— amplitude shift keying ASK

— on-off keying OOK

— differential phase shift keying DPSK

— spread spectrum pseudo-random noise SS.

6. INFERENCE IN COMMUNICATION NETWORKS

6.1. Following now classical failure diagnosis methodology {42,44}, we decompose the communication system diagnostic inference problem into the following steps and knowledge subcategories (as done in {43}) {see Fig.2}:

1. Specifications of failure and event modes: (E)

— Hardware failure or degradations at information source, destination, or during transmission (nodes links)

— Software failures, at same locations

— Procedural failures:
  o Information input errors
  o Information transmission errors (flow control, routing, protocol, transactions)
  o Information output/display errors
  o Information analysis errors
  o Action errors
  o Security procedures

— Human failures

— Systems operations

— Decision-making failures.

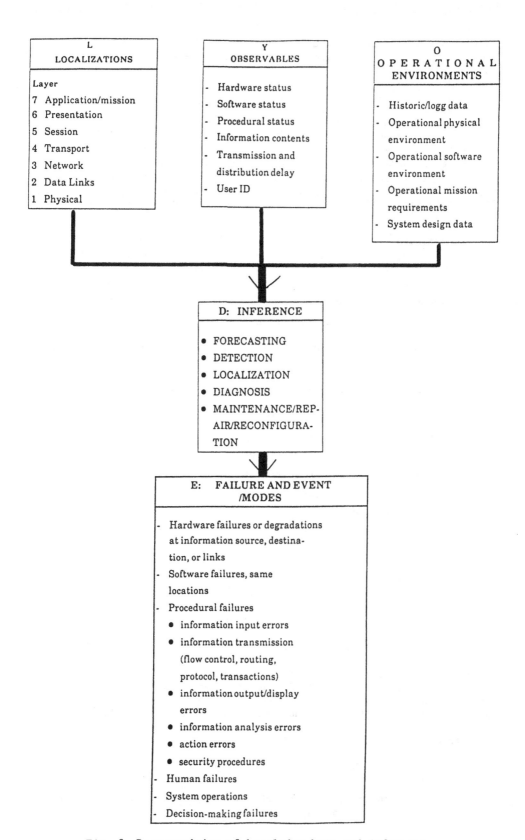

Fig. 2 Decomposition of knowledge base and inference.

ii. Failure localizations (L). We suggest to take as generic failure localizations the seven layers of the OSI architecture for open communications networks {45}:

Layer 7: Application/mission
Layer 6: Presentation
Layer 5: Session
Layer 4: Transport
Layer 3: Network
Layer 2: Data links
Layer 1: Physical

iii. Observables (Y) {42,44}. Among the possible observables, we include for each localization:

— Hardware status

— Software status

— Procedural status

— Information contents

— Information transmission and distribution.

These observables include the test and monitoring features of Section 2. (types A, D, T).

iv. Operational environment (O)

This covers exogeneous conditions:
— Historic data {42}
— Operational physical environment
— Operational software environment
— Operational mission specifications/requirements
— System design data

v. Inference tasks (I)

o Forecasting
o Detection
o Localization
o Diagnosis
o Maintenance/repair/reconfiguaration

Making a diagnostic inference in communications systems in then specifying fully the mappings d {42,44}:
$d \in I: (Y \times L \times O) \to E$, for a given application. We will throughtout use the above framework, as the left hand side in d specifies the knowledge bases required, and the right hand side tells the goals to be inferred. Elements in Y are acquired by the test and monitoring units (see Sections 3,4).

6.2. The contributions of artificial intelligence {34} reside in:

— selection of knowledge representations able to handle jointly elements of $Y \times L \times O$

— elicitation/acquisition of additional knowledge about Y,L,O,E, in the form of knowledge bases

— inferencing d ∈ I and e ∈ E from current information about Y x L x O, supplemented by knowledge bases about the same

— offering interactive man-machine facilities in carrying out and explaining the inference d, while reducing operator work-load.

In the remaining Sections 7-11, the paragraph entitled "Framework" highlights which knowledge bases are used for each generic application.

7. KNOWLEDGE BASED FAILURE DETECTION AND DIAGNOSIS AT THE PHYSICAL LEVEL

7.1. Framework

E: Hardware failures, and procedural failures
L: Physical level, and data links (layers 1 and 2)
Y: Hardware status, procedural status
O: Physical environment, mission requirements, design data
I: Detection, localization, diagnosis, repair.

7.2. Approach

Knowledge based failure detection and diagnostic systems (also called diagnostic expert systems) {44} rely on distributed specialized knowledge bases about elements of E, L, Y, O, with possibly a common blackboard architecture, to perform the tasts d ∈ I using heuristic search, test generation, forward/backward/mixed chaining, constraint propagation, and diagnostic metarules.

Three major difficulties exist:

i knowledge acquisition about failure/degradation modes {42}

ii selection of useful sensor data and observables, by feature extraction {42}, although possible features are already identified (see Section 1)

iii still today insufficient validation of the completeness of the knowledge bases, in view of safety, repeatability, and reduced diagnostic requirements.

Four major contributions are expected, or already experienced, from knowledge engineering techniques {44}:

1) generation of functional test sequences through metarules;

2) reduction of false alarms through in-built test (BIT) with knowledge bases {39},{24 };

3) reduction of skills in test and maintenance tasks

4) use of diverse observation types (see also Sections 1 and 10)

These contributions are likely to go into seven generic systems:

a) computed aided preliminary design for testability {46}

b) smart built-in-test to identify intermittent faults, reduce false alarms, and carry out recalibration {42}

c) smart system integrated test for system level testing while in operation {42}

10. SECURITY SCREENS

10.1. Framework

    E: Security procedures, information input errors, information output
       errors
    L: Session, or applications (Layers 5,7)
    Y: Procedural status, information contents
    O: Operational software environment
    I: Failure/degradation detection, reconfiguration.

10.2. Approach

As described already in Section 8, natural language analysis of the
messages may generate explicit security levels and distribution screens
as encoded in the messages.

However, knowledge based approaches may also be considered applied
to the user or operator of the communications network, in order to control their access to the information conveyed and to network controls.
Through a dialogue with the KBS system, the user or operator will have to
establish their clearances and need to know.

Specific cases hereof are:

— knowledge based access to a multi-level security network controller, through the protocol and distributed operating system

— knowledge based system in the analysis of key management systems

— global authentification service or key distribution service,
  making use of trust relationship between network users and of
  composition between channels {22}.

The limitations are:

— lack of exhaustive knowledge about screen avoidance, such as
  spoofing

— insufficient development of truth maintenance techniques, to
  establish the relation between candidate keys and fixed procedures.

11. NETWORK OPERATIONS

11.1. Framework

    E: Hardware failures, information transmission errors, system opera-
       tions
    L: Network, transport (Layers 3,4)
    Y: Hardware status, information transmission and distribution delay
    O: Operational physical environment, operational mission specifica-
       tions (requirements).
    I: Reconfiguration.

11.2. Approach

The goal is to maximize an economic criterion related to network rev-
enues, by having a knowledge based system assist the operator in capacity
allocation and planning, incl. flow control {24}.

Examples are:

— capacity assignment in networks, based on network monitoring (especially of traffic measurements T of Section 2)

— network status intelligent user interface; this could be an object or window front-end, where objects/windows pop up depending on events and recommended corrective actions (change of routing or capacity; status information (time, traffic intensities, costs, data flow topology)) are displayed continuously {24}.

The limitations are mostly in terms of limiting the consequences of human decision making errors, due to insufficient knowledge based systems validation.

12. QUEUE OBJECTS

12.1. Framework

E: Information transmission errors, systems operations
L: Data links, network (Layers 2,3)
Y: Software status
O: Operational software environment
I: Reconfiguration, localization.

12.2. Approach

Artificial intelligence software development environments, especially object oriented ones (SMALLTALK; PROLOG-S, etc.), allow for the definition of queue objects to describe queuing on a node or link. The introduction of queue objects allows the operations and permissible network contexts of queues to be precisely controlled.

A queue declaration could be the frame:

```
queue-declaration      ::=      queue-identifier list:
      subtype-indication (queue-specification);

queue-specification    ::=      size specification    |
                                file-association

size-specification     ::=      size static-expression;

file-association       ::=      file (direction)
                                file-simple-name;

direction    ::=  in  |  out        (default:in);
```

The subtype indication defines the subtypes of the elements of the queue. A queue specification is exactly one of a size specification on the queue, or a binding of the queue to a file. The queue objects allow to define consistently interface lists and file associations.

The four interactions between queues and a communication process are four objects: Send, Receive, Is Empty, Is Full.

Each Send call to a queue puts the value into the queue at the end of the queue; the queue can contain either an indefinite number of objects (if no size specification appears in the queue declaration) or a definite number (the number declared as part of the size specification). In the case of a queue with a size specification, the queue can hold only that many values; Is Full is true exactly when the queue contains that many values. Is Full is always false for queues of indefinite size, which includes queues associated with files. A Send done on a queue for which Is Full is true causes the process containing the call to suspend until a Receive is done on the queue.

A call to Receive on a queue removes a value from the beginning of the queue. Is Empty is true when the queue contains no values. A Receive attempted on a queue for which Is Empty is true and for which there is no file association suspends the process containing the call until a Send is done on the queue. A Receive attempted on a queue for which Is Empty is true and for which there is a file association either causes indefinite suspension of the process containing the call or it is an error.

Both Send and Receive cause the process containing the call to be suspended for a minimum of one cycle. Sends and Receive are always processed in the order issued.

There is an inherent non-determinism in the queue mechanism that becomes apparent when more than one process sends to or receives from a queue. This non-determinism, results from the lack of a defined order of process execution, and is well-behaved, in the sense that specific processes execute in the same order repeatedly when using a given procedure, although they may execute in a different order when using other procedures. This is especially useful when carrying search and inference by a knowledge based procedure, as the above queue object definition will intrinsically respect the information distribution procedures above.

REFERENCES

1. W. J. Garner, Bit error probabilities relates to data link S/N, *Microwaves*, Vol 17, no 11, 101-105, (November 1978).

2. L. F. Pau, Specification of an automated test system vs. optimum maintenance policy and equipment reliability, Proc. 1979 Ann. reliability and maintainability symposium, Washington DC, 23-25, IEEE Cat. 79 CH 1429 - OR, 147-152 (January 1979).

3. L. F. Pau, *Failure diagnosis and performance monitoring*, Marcel Dekker, Publ., New York, (1981).

4. W. Boghdady, Flow control with a failure detection and localization capability in high data rate transmission networks, T. R. Enst, Dept. Electronics, E.N.S. Télécommunications, Paris, (June 1979).

5. A. Chatterjee, N. D. Georganao, Analysis of a packet switched network with end-to-end congestion control and random routing; *IEEE trans. Comm.*, Vol COM - 25. no 12, 1485-1489 (December 1977).

6. M. Meri, A reliable control protocol for high-speed packet transmission, *IEEE Trans. Comm.*, Vol COM - 25, no 10, 1203-1209 (October 1977).

7. J. Labetoulle, G. Pujolle, Modelling and performance evaluation of the HDLC protocol, Int. Symp. on flow control in data networks, IRIA, Rocquencourt, Paper 8.2 (February 1979).

8. W. W. Chu, Optimal message block size for computer communications with error detection and retransmission strategies, *IEEE Trans. Comm.*, Vol COM-22, No 10, 1516-1524 (October 1974).

9. M. Robin, Contrôle optimal de files d'attente, T. R. 117, IRIA, Rocquencourt, (May 1975).

10. T. Collings, C. Stroneman, The M/M/∞ queue with varying arrival and departure rates, *Operations Research*, Vol 24, no 4, 760-773 (July-August 1976).

11. J. Labetoulle, G. Pujolle, Modelling of packet switching networks with finite buffer size at each node, T. R. 239. IRIA, Rocquencourt, (June 1977).

12. M. Dal Cin, Performance evaluation of self-diagnosing multiprocessing systems, Proc. FTCS-8 Toulouse, IEEE Cat 78.CH 1286 - 4C, 59 -64 (June 1978).

13. B. Goldman, Deablock detection in computer networks, AD - A - 047025, p. 180 (1977).

14. D. Towsley, Error detection and retransmission strategies in computer communication networks, Proc. 1978 COMPCON Fall, Washington, IEEE Cat 78 CH 1388 - 8C, 12-18 (September 1978).

15. A. Segall, P. M. Merlin, R. G. Gallager, A recoverable protocol for loop-free distribution routing, 1978 Int. Conf. Communications, Toronto, IEEE Cat 78 CH 1350 - 8 CSCB, Paper 3.5 (June 1978).

16. V. Baek Iversen, Review of traffic measurement methods and application fields of measurement results, T. R. 18-78, IMSOR, Technical Univ. of Denmark, Lyngby, (1978).

17. G. Gardarin, B. Piot, Détection des interférences er évaluation de leur fréquence, IRIA Grant 77.007, Inst. Programmation, Univ. Paris (June 1978).

18. J. Clavier, N. Niquil, G. Coffinet, F. Behr, *Théorie et technique de la transmission de données*, Masson, Paris, (2nd Edition) (1977).

19. R. van Slyke, H. Frank, Network reliability analysis, *Network*, Vol 1, (1972).

20. L. Fratta, U. Montanari, Decomposition techniques for evaluation network reliability, Int. Teletraffic congress - 8, Melbourne (1976).

21. L. Svoboda, Performance problems in distributed systems, *INFOR*, Vol 18, No 1, 21-40 (February 1980).

22. A. D. Friedman, L. Simoncini, System level fault diagnosis, *IEEE Computer*, 47-53 (March 1980).

23. R. Freiburger, *Statistical computer performance evaluation*, Academic Press, NY, (1971).

24. L. F. Pau, W. Boghdady, A. Bousquet, Fault detection capability implications for flow control in data communications networks, Proc. AUTOTESTCON 1979, Mineapolis, Publ. by IEEE, (October 1979).

25. E. Rasel, Generating error correction codes, *Systems International*, 32-36 (July 1979).

26. L. Meeks, Characterization of instrumentation tape signal dropouts for approapriate error correction strategies in high density degital

recording systems, Honeywell, Test Instruments Div., PoBox 5227, Denver, CO 80217.

27. R. Johannesson, E. Paaske, Further results on binary convolutional codes with an optimum distance profile, *IEEE Trans. Inform. Theory,* Vol IT - 24, no 2, 264-268 (March 1978).

28. E. Paaske, Short binary convolutional codes with maximal distance for rates 2/3 and 3/4, *IEEE Trans. Inform. Theory,* Vol IT - 20, no 5, 683-689 (September 1974).

29. G. F. Erbrecht, Automated field measurement system for T-1 characterization, ICC'80, IEEE Cat. 80 CH 1505 - 6 CSCB.

30. T. C. Knapp, Smart - 1: a systematic monitor and remoter of T-information, ICC'80, See {29}.

31. D. W. Jurling, T-carrior performance, ICC'80, See {29}.

32. S. Takenaka et al, Bit error rate monitor for 4 phase PSK system, ICC'80, See {29}.

33. Command, Control and Communications, *Aviation Week and Space Technology,* (December 1985).

34. A. Barr, E. Feigenbaum, (Ed.), *Handbook of Artificial Intelligence,* William Kaufman, San Francisco (1982).

35. M. Athans, The expert team of experts approach to command and control organizations, *Control Systems Magazine,* (September 1982).

36. J. Froscher, R. Grishman, a linguistically motivated approach to automated analysis of military messages, Proc. 1983 American Artificial Intelligence Conf., Rochester, MI, (1983).

37. Expert systems in government, Proc. IEEE 85 CH-2225-1, Washington, IEEE Cat. 85 CH-2225-1 (October 1985).

38. L. F. Pau, J. M. Negret, Intelligent contents based message routing in communications networks, Proc. Nordic Teletraffic Conference, Lyngby, Denmark, (August 1986).

39. L. F. Pau, Data Communications network monitoring for failure diagnosis in the presence of non-digital links, T. R., E.N.S.-Télécommunications -D-80008, Paris, (1980).

40. L. Kleinrock, Distributed systems, *IEEE Computer Magazine,* (November 1985).

41. L. F. Pau, (Ed.), *Artificial Intelligence in Economics and Management,* North Holland, Amsterdam, (1986).

42. L. F. Pau, *Failure Diagnosis and Performance Monitoring,* Marcel Dekker Publ., NY, (1981).

43. L. F. Pau, J. M. Negret, SOFTM, a software mmintenance expert system, Internal report, Battelle Memorial Institute, (September 1985).

44. L. F. Pau, Survey of expert systems for fault detection, test generation and maintenance, *Expert Systems J.,* vol 3, no 2, 100-111 (April 1986).

45. C. L. Heitmeyer, S. H. Wilson, Military message systems: current status and future directions, *IEEE Trans. Communications*, Vol COM-28, no 9, (September 1980).

46. Special issue on artificial intelligence techniques, *IEEE J, Design and Test of Computers*, Vol 2, no 4, (August 1985).

47. D. E. Bell, L. I. La Padula, Secure Computer Systems, Vol. I-II, MTR 2547, MITRE Corp., Bedford, (November 1973).

48. L. F. Pau, B. Wafae, A. Bousquet, Fault detection capability implications for flow control in data communication networks, Proc. 1979 IEEE Int. Conf. AUTOTESTCON, Minneapolis, (September 1979).

49. L. Bolc (Ed.), *Natural Language Communication with Computers*, Springer Verlag, NY, (1978).

50. L. F. Pau, An expert system Kernel for the analysis of strategies over time in {41}, 107

51. L. F. Pau, Fusion of multisensor data in pattern recognition, in J. Kittler, K. S. Fu, L. F. Pau (Ed.), *Pattern Recognition Theory and Applications*, D. Reidel Publ. Co, Boston (1981).

52. L. F. Pau, Inference of the structure of economic reasoning from natural language analysis, *J. Decision Support Systems*, Vol 1, 313-321 (1985).

53. B. J. Grosz, K. Sparck Jones, B. L. Webber (Ed.), *Readings in Natural Language Processing*, Morgan Kaufman Publ., Los Altos (CA), (1986).

54. A. D. Birrell, B. W. Lampson, R. M. Needham, M. D. Schroeder, A global authentification service without global trust, Proc. IEEE Symposium on security and privacy, Oakland, 223-230 (April 1986).

55. Use of learning algorithms in telephone traffic routing, *Automatica*, Vol 19, no 5, 495-502.

56. *IEEE Communications Magazine*, Vol 24, no 1, 28 (January 1986).

56. G. Hirst, *Semantic Interpretation and the Resolution of Ambiguity*, Cambridge Univ. Press (1987).

16

A SURVEY OF EXPERT SYSTEMS FOR EQUIPMENT MAINTENANCE AND DIAGNOSTICS

William T. Scherer and Chelsea C. White, III

University of Virginia
School of Engineering and Applied Science
Department of Systems Engineering
Thornton Hall
Charlottesville, VA 22901 USA

1. INTRODUCTION

Considerable research is being conducted in the area of expert systems for diagnosis. Early work was concentrated in medical diagnostic systems (Clancey and Shortliffe, {21}). MYCIN (Buchanan and Shortliffe, {15}) appears to represent the first medical diagnostic expert system. Current efforts are expanding to the area of equipment maintenance and diagnostics, with numerous systems having been built during the past several years. We concentrate our effort in this survey on expert systems for diagnosis and refer the reader to (Hayes-Roth, Waterman, and Lenat, {42}; Hayes-Roth, {43}), (Waterman, {101}), and (Charniak and McDermott, {19}) for introductions to expert systems. We remark that it is common for diagnostic systems to integrate concepts from artificial intelligence (expert systems), decision theory, and operations research.

2. EXPERT SYSTEMS: GENERAL

The two basic components of an expert system (ES) are the *knowledge base* (KB), which consists of factual knowledge and preferential knowledge, and the *inference engine*, which directs (controls) the use of the KB. Numerous supporting components (e.g., the user interface) may also be incorporated into an ES. Various types of knowledge representation schemes can be employed: frames, production rules, lists of facts, semantic networks, logic predicates, etc. The complexity of most diagnostic systems makes knowledge representation a critical issue; see (Pau, {71}) for an evaluation of knowledge representation schemes; We note, however, that an ES will often require a combination of representations. With regard to the inference mechanism for ESs, the various options include: forward-backward chaining (still the most common), generate/test methods, heuristic search, and meta-rules (Pau, {71}). Combinations of the above inference mechanisms are also common.

Three basic types of ESs, in order of increasing generality, are production systems, structured production systems, and distributed systems (Pau, {71}). Production systems have a single KB and a single inference engine, while structured production systems use meta-rules to move between several KBs. Distributed systems use a hierarchically structured network of cooperating specialists (Bylander, Mittal, and Chandrasekaran, {16}). Distributed systems are the most general of the three.

3. EXPERT SYSTEMS: DIAGNOSTIC

The first diagnostic expert systems (DESs) for fault diagnosis were developed in the early 1970's at MIT. These systems were unique in that they modelled the normal behavior and structure of the system rather than a tree of possible faults. An important feature of these systems was their explanation capabilities, usually generated via a backtrace of the analysis. The early systems such as EL, LOCAL, and DESI viewed the device as a collection of components with given component interactions, where the components were organized in a hierarchical fashion (King, {53}). These systems were often based on a circuit-level description of the device. The DART project (Genesereth, {36}), however, approached the diagnosis problem from a different perspective. Using the EMYCIN framework for medical diagnosis, DART is a system for diagnosing IBM-370 class computers that is not based on the circuit level but on a system level. The philosophical basis for this approach, which has become termed "shallow-reasoning", came from the medical diagnostic systems, where a complete description at the "circuit" level is not fully understood or is highly complex. Often in medical diagnosis, rules are used that relate sets of symptoms to possible diseases. These "pattern->action" rules are based on problem-solving heuristics generated by the expert and often employ a simple control structure in using the knowledge-base (Chandrasekaran, et al., {18}). This is in contrast to the early DESs, where the model was one of normal functioning. It is also in contrast to the diagnostic systems that employ the notion of "deep-reasoning" to determine faults (see Davis, {22}; Pan, {70}). Deep-reasoning involves using lower-level reasoning on causal, functional, and spatial information, while shallow reasoning refers to the use of rules-of-thumb that are often acquired from experience. In many instances, however, it is difficult and too complex to define the interactions at the system level.

Conventional fault diagnosis, used by reliability engineers, employs fault trees and their successors, cause-consequence trees (CCT). The CCT methodology models faults that can arise in a system, along with associated logical, causal, and temporal relationships. CCTs require a significant level of effort in their construction, e.g., the construction of a CCT for a nuclear plant required approximately ten man-years (King, {53}). Major difficulties with CCT models is that they do not allow feedback and a priori information regarding the fault is difficult to take advantage of effectively. Weaknesses in AI based techniques, as compared to CCT, include: the requirement of a complete model of the device, the assumption of reliable measurements, the lack of guidance with respect to what course of action to take other than which component to replace, and the lack of time-critical processing (King, {53}). We note that many AI based expert systems for diagnosis employ structures and techniques from conventional fault diagnosis.

Experienced diagnosticians use a combination of techniques for diagnosis, including a familiarity with the system documentation, a functional understanding of the system components, an understanding of the system interrelationships, knowledge of the failure history of the device, along with numerous heuristics (Pau, {71}). It is this complex mixture of knowledge that complicates the diagnostic problem. Medical diagnosis and treatment expert systems are often hindered by the lack of a complete understanding of the human body, while equipment diagnostic systems can have available a complete range of information. Such information includes a complete schematic diagram of all circuits involved, functional models of behavior, spatial models of behavior, and reasoning by second principles. This rich amount of information allows for expert diagnostic systems of potentially high accuracy and considerable complexity.

In current AI-based diagnostic systems there are several common notions and concepts. The diagnostic problem is often viewed as a discrepancy between predicted and observed behavior. Systems such as DART, however, often relate observed behavior to faults from a system viewpoint. Also, diagnostic systems attempt to exploit hierarchical behavior whenever possible. This allows a modular system where the degree of detail associated with any module within any level is not required to be consistent. The interactions between modules and levels is kept as simple as possible.

Critical issues in expert system design for electronic equipment diagnosis, therefore, are:1) the level at which to approach the design (e.g., at the schematic diagram level or the functional model level), 2) the knowledge representation scheme to be used once the level has been selected (e.g., IF-THEN rules, frames, etc.), and 3) the method by which uncertainty is to be handled (e.g., probabilities, certainty factors, fuzzy sets, etc.). Important in the knowledge representation scheme is the notion of reasoning from first principles, which relates to the level at which the system being diagnosed is described (Reiter, {80}). Davis {23} argues for "models of causal interaction" rather than the traditional models of fault diagnosis. Often, an expert diagnostician is required to revert to "deep" knowledge when confronted with a particularly difficult problem or when required to explain their diagnostic process (Chandrasekaran and Mittal, {18}; Davis, {23}; Herbert and Williams, {45}; Downing, {28}; Scarl, Jamieson, and Delaune, {87}). Also with regard to knowledge representation, the concept of functional versus behavioral models is an important distinction. A diagnostic system can model a component of the system with only the notion of the behavior of the component without regard to function. This permits the behavior of each component to be understood in a context-free manner, which allows for relatively easy interchange of modules in a hierarchical system. (See Milne, {60}, for a detailed discussion of system architectures). Uncertainty, one of the most important issues in DESs, is discussed in detail in the following section.

4. UNCERTAINTY IN DES

Diagnostic expert systems can be viewed as a search for a single fault or multiple faults, where the search is usually heuristic in nature. Inherent in most diagnostic systems is the notion of uncertainty, i.e., the knowledge the system has concerning the cause of the fault is not known exactly. Early DESs employed IF-THEN rules and a simple chaining mechanism as the knowledge base and inference engine (control structure), respectively. Current expert diagnostic systems are employing sophisticated techniques to deal with the issue of uncertainty. The reader is referred to the Proceedings of Uncertainty in Artificial Intelligence from the recent AAAI conference for an extensive collection of current research papers dealing with uncertainty in AI. Other surveys of uncertainty in diagnostic systems include (Dutta, {30}), (Kanal and Lemmer, {52}), and (Goodman and Nguyen, {37}).

Five general (non-exhaustive) techniques exist for dealing with uncertainty: 1) Probabilities, 2) Fuzzy Sets, 3) Certainty Factors, 4) Dempster-Shafer Theory, and 5) Set-Covering Techniques. These are general categories and most systems use a modified version of one of the above. We refer the reader to (Kanal and Lemmer, {52}) for a description of the above uncertainty models.

Dempster-Shafer theory is a generalization of Bayesian theory that is receiving growing interest from the developers. For an introduction to Dempster-Shafer theory see (Shafer, {90}) and (Dempster, {27}). For example, Cantone and colleagues {17} describe a diagnostic system that uses probabilistic search for electronics troubleshooting. Their system does not require the probabilities to sum to one and updating is done via Dempster's rule. However, these probabilities have values between 0 and 1 and have a relative frequency interpretation. Two kinds of rule updating are allowed: evidence that is independent and combined with earlier evidence, and non-independent evidence that revises belief about other components in the system. Recent work on Dempster-Shafer theory includes (Zadeh, {109}), (Biswas and Anand, {11}), (Rowe, and colleagues {85}) and (Yen, {107}).

Fuzzy sets (Zadeh, {108}) are receiving considerable attention, much of which is in the medical diagnostic system literature (Bandler and Kohout, {6}; Bolc and colleagues, {12}; Gazdik, {35}; Moon and colleagues, {63}; Negotia, {65}; Ralescu, {79}; Turksen, {99}; Umeyama, {100}; Kacprzyk and Yager, {50}; Baldwin, {5}; Kitowski and Ksiazek, {54}). Advantages of fuzzy sets for medical diagnosis (and general diagnosis) include the ability to define inexact entities as fuzzy sets, the idea that fuzzy sets provide a linguistic approach that closely approximates medical texts, and the ability of fuzzy sets to draw approximate inferences (Adlassnig, {2}).

Probably the most common uncertainty description is the use of certainty factors associated with rules in the rule base. Certainty factor systems update beliefs by using heuristic indices, often with questionable assumptions (Yadrick and colleagues, {106}). First described in the MYCIN project (Buchanan and Shortliffe, {15}) and in the PROSPECTOR system (Yadrick and colleagues, {106}; Duda, Gaschnig, and Hart, {29}), numerous systems employ a modified form of certainty factors (Odryna and Strajwas {68}; Merry, {59}). Recently there has been considerable interest regarding the relationship between belief updating and probabilities (Heckerman, {44}).

Set-covering models for diagnostic problem solving are also receiving current attention (Reggia, Nau, and Wang, {81-82}; Reggia, Na Wang and Peng, {83}; Kalme, {51}; de Kleer and Williams, {26}; Reiter, {80}). Major features of the set-covering model include the adaptability of the model to more general problems and the ability of the model to handle multiple faults. Set-covering models attempt to explain all the symptoms with a minimum of complexity, e.g., the minimum number of faults (Reggia, Nan and Wang, {81}). INTERNIST (Masarie, Miller, and Myers, {57}; Miller, Pople, and Myers {61}) is a system that is closest to a set-covering model. Recent work on set-covering models involves incorporating probabilistic inference and symbolic interence (Peng and Reggia, {75-76}).

One of the earliest mechanisms for dealing with uncertainty in diagnostic expert sykstems is Bayesian based probabilistic reasoning (Clancy and Shortliffe, {21}). In the 1960s many medical diagnostic systems were developed that employed a Bayesian framework. Several difficulties with Bayesian based systems brought about the decline in their use (Shortliffe, Buchanan, and Feigenbaum, {91}). Major difficulties with Bayesian systems include: 1) the lack of generality, 2) the extensive data requirements if independence and mutual exclusiveness are not assumed, 3) the requirement of precise probabil-ities, 4) lack of an explanation capability and 5) the problematic acceptability of the systems. Despite these difficulties, Bayesian systems that have been developed often have excellent performance (Clancey and Shortliffe, {21}). Recent efforts to address the issues involved with Bayesian systems include (White, {102}), (Snow, {93}), (Shachter and Bertrand, {89}), (Pearl, {74}). (Norton, {67}), (Barth and Norton, {7}) and (Schwartz, Baron, and Clark, {88}).

5. DEVELOPMENT ENVIRONMENTS

Many diverse environments are being used for the development and implementation of expert systems. We noted based on this survey that the language and environment selected is usually based on availability of the hardware, the software (language and environment), and the expertise. It is also common to develop the system in one environment and then to transfer it to another environment for actual implementation, e.g., PUFF and AEGIS (Adams and colleagues, {1}; Aikins, Kunz, and Shortliffe, {4}).

Three basic approaches can be used when selecting software for expert diagnostic system development. The first is to design the system from a fundamentally low-level language, such as LISP, PROLOG, or C (or BASIC, PASCAL or FORTRAN). The second approach is to use an epxert system shell program or expert system language that usually has a built-in knowledge base form and a relatively fixed control structure. Examples of such systems include EXPERT, KEE, and OPS5 (Brownston and colleagues, {14}). There are also expert system shells designed exclusively for building diagnostic systems (see "shells" in the sample system list below). See (Waterman, {101}) for a description of numerous general purpose expert system shells. The third approach is a combination of the first two - modifying a shell system for a particular application. This approach is advantageous, if the shell system allows user modification, in that the designer saves considerable development time by using as much of the existing structure as possible.

In the systems described in the applications section below, one of the most common software environments used was the OPS5 ES language (IDT, ANGY, PROD, Toast, The Diagnostician, MELD), while the EXPERT shell system was also common (AEGIS, APRES). Other shell systems used were S.1 (Mark 45 FDA), SAGE (AMF), and Inference ART (APRES). Languages used include FORTH (DELTA), FORTRAN (EXPERT), LISP (IN-ATE, ACES, DORIS, FIS), PROLOG (DIGS, APEX 3), and PASCAL (CATA). Several of the systems were based on or similar to MYCIN or EMYCIN (DART, PUFF, PDS). There is a current debate as to whether LISP will become popular as an implementation language or tend to remain as a development language (e.g., Jacobson, {48}; Gabriel, {34}). Currently, there is also considerable interest in the language C for development and implementation. With respect to hardware, there was a wide variety of machines used, with the most common being the DEC VAX computers. Also used was the DEC PDP, dedicated LISP machines (e.g., Symbolics, TI Explorer), and various microcomputers.

6. APPLICATIONS AREAS

Application areas include: computer malfunction diagnosis (DART, IDT; Agre and colleagues, {3}), power networks (TOAST; Liu and Tomsovic, {56}; Zwingelstein and Monnier, {110}), faults in VLSI chips (PROD, MIND), chemical plant fault diagnosis (Palowitch and Kramer, {69}), communications networks (Williams, {105}), ATE diagnosis (Ryan and

Wilkinson, {86}), software diagnosis (Jackson, {47}), aircraft diagnostics (Montgomery, {62}), and manufacturing cells (Chiu and Niedermeyer, {20}).

Following are some sample DESs, followed by sample DES shell systems. The list is by no means exhaustive, and the brief descriptions present information that we feel is of interest. We remark that it is often difficult to determine whether a system is in actual use, in the developmental stages, or a paper model only.

Sample systems:

ACES: (Attitude Control Expert System) Performs diagnosis of failures in the attitude control system of the DSCS-III satellite. Uses failure-driven learning to improve diagnostic capability. System implemented in a combination of LISP and PROLOG on a Symbolics 3600. Developed by the Aerospace Corporation. (*Pazzani, 1986*).

AEGIS EXPERT: Built using the EXPERT shell development system (developed at Rutgers University for medical diagnosis). System adds diagnostic capabilities to the shipboard AEGIS system. Uses IF-THEN rules. The final version will be translated into C to run on an IBM AT. Developed at Advanced Technology Laboratories. (*Adams and colleagues, 1986*).

AMF: (Advanced Maintenance Facility) A DES for TXE4A telephone exchanges (England). The system has approximately 1350 rules, 2200 questions, 2400 assertions, and 2300 items of advice, arranged into a net with 660 areas relating to specific fault indications. Developed using the SAGE expert system shell on a 16-bit microcomputer running under UNIX. (*Thandasseri, 1986*).

APRES: Performs an analysis of computer crash dumps in order to determine computer hardware faults. Originally coded in the EXPERT shell system, it has been translated into Inference ART, which resides on a VAX in a LISP environment. The system has approximately 1000 rules using the EXPERT shell system. (*Jackson, 1985*).

CATA: (Computer-Aided Test Analysis System) Helps engineers design electronic systems for testability. Written in PASCAL on a VAX 11-780. (*Robach, Malecha, and Michel, 1984*).

CRIB : (Computer Retrieval Incidence Bank) Designed for computer fault diagnosis using a pattern-directed inference system. (*Hartley, 1984*).

DART: (Diagnostic Assistance Reference Tool) Assists a technician in finding faults in a computer system. The first version was EMYCIN based, while the second version used meta-level reasoning system (MRS). (*Genesereth, 1984; Bennett and Hollander, 1981*).

DELTA: (Diesel Electric Locomotive Troubleshooting Aid, field prototype is CATS-1, for Computer-Aided Troubleshooting System) Implemented in FORTH on a DEC PDP 11/23. Has approximately 530 rules and uses a mixed configuration inference engine that has forward chaining and backward chaining. (*Bonissone and Johnson, 1983*).

The Diagnostician: (see "Toast") Diagnoses faults in a power network (Allegheny Power System). Written in OPS5, the system has approximately 350 rules. Designed for a Distributed Problem Solving (DSP) environment, where there are human problem solvers that interact with a number of programmed problem solvers. (*Talukdar, Cardozo, and Perry, 1986*).

DIEX (Diagnostic Expert System): Diagnoses chemical plant operations. Divides problem into subtasks, where different strategies are used for each subtask, including pattern-matching and model-based algorithms. (*Palowitch and Kramer, 1985*).

DORIS (Diagnostic Oriented Rockwell Intelligent System): A rule based system for aiding in fault diagnosis. Allows forward and backward chaining, and has a rule maintenance system and an explanation system. Various versions are coded in InterLisp on a VAX 11/750, Lisp-88 on an IBM PC, and ZetaLisp on a Symbolics. *(Davis, 1986)*.

ET: (Expert Technician, Experience Trapper) Probabilistic, heuristic based diagnostic system for PRODUCT, a complex electronic device manufactured by Texas Instruments. Uses weighted predictive functions to rank actions that will remove symptoms. Also employs machine learning and adoption techniques to modify and add knowledge concerning its domain. Employs an extended version of the Dempster-Shafer theory of evidential reasoning. *(Rowe, 1986)*.

Fieldserve: (Field Service Advisor) Diagnoses electromechanical systems, specifically in the field. Runs on a Vax 11/780 in Franz LISP and the general purpose inference engine Genie (Vanderbilt). Frame based, employs IF-THEN rules. *(Hofmann and colleagues, 1986)*.

FIS: (Fault Isolation System) Uses a rule-based causal model to update beliefs and recommend tests for electronic equipment. Information provided in set description and probabilities. Written in Franz LISP on VAX 11/780, developed at US Naval Research Lab. *(Pipitone, 1986)*.

Hitest: Knowledge-based automatic test generation system for digital circuits. *(Bending, 1984)*.

IDT: (Intelligent Diagnostic Tool): Identifies faults in PDP 11/03 computers. Uses OPS5 and was developed at Carnegie-Mellon University. *(Shubin and Ulich, 1982)*.

IN-ATE: Electronic troubleshooting aid designed to guide a novice technician. IN-ATE uses probabilistic search to determine faults. The search is guided by three parameters: test cost, conditional probabilities of test outcomes, and proximity to a solution. Written in Franz LISP on a VAX 11/780. *(Cantone and colleagues, 1983)*.

LES (Lockheed Expert System): A system for diagnosing a large signal-switching network. Uses production rules and frames to describe least repairable units (LRUs). A goal-driven, backward chaining system, LES also allows for data-driven, forward chaining reasoning. Has approximately 70 rules and 1200 facts. *(Laffey, Perkins, and Nguyen, 1984)*.

MARK 45 FDA (Fault Diagnosis Advisor): An integrated expert system for the diagnosis and repair of the Mark 45 Naval Gunmount. Implemented in the S.1 expert system programming language system developed by Teknowledge, Inc., which is a descendent of EMYCIN. Integrates maintenance history database, spare parts inventory database, BIT, ATE, and other KBS. *(Powell, Pickering, and Westcourt, 1986)*.

MDX: Decomposes diagnostic knowledge into a classification hierarchy of cooperating specialists. Uses an "establish-refine" mechanism to control movement through the hierarchy. *(Sticklen and colleagues, 1984; Chandrasekaran and Mittal, 1984)*.

MELD (Meta-Level Diagnosis): Diagnosis of electromechanical systems. Two levels of reasoning are used - object-level and meta-level. Written in OPS5. *(Thompson and Wojcik, 1984)*.

MIND: (Machine for Intelligent Diagnosis) An expert system for VLSI test system diagnosis. Hierarchically structured rule-based system designed to reduce the MTTR. *(Wilkinson, 1985) (Ryan and Wilkinson, 1986) (Grillmeyer and Wilkinson, 1985)*.

NDS: (Network Diagnostic System) An expert system for fault isolation in a nationwide communications network (COMNET). Based on ARBY (McDermott,

1982), the system performs a heuristic search for multiple faults through a space of hypotheses. NDS has approximately 150 ARBY rules and the performance level is that of an intermediate level diagnostician. (*Williams, 1983*).

PROD: A complete diagnostic system for determining faults in VLSI chips. Matches measurements on chip parameters to a library of faults. A production rule system written in OPS5 then generates a set of certainty factors for the most likely faults. Main system implemented using a UNIX file system. (*Odryna and Strojwas, 1985*).

PUFF: Highly successful system for interpreting pulmonary function test results. Originally developed using EMYCIN on Stanford's SUMEX-AIM computer, it has been translated into BASIC on a PDP-11 for on-site use. (*Aikins, Kunz, Shortliffe, 1983*).

REACTOR: An expert system for the diagnosis of nuclear reactor accidents. (*Nelson, 1982*).

Toast: (see "The Diagnostician") An expert system for simulating events in power networks and diagnosing problems. Uses Cops (super-set of OPS5) on a network of VAX computers. (*Talukdar, Cardozo, and Leao, 1986a*).

Sample Diagnostic Shell Systems:

APEX 3: An expert system shell for fault diagnosis. Uses forward and backward chaining as its main control mechanism. Employs the uncertainty system developed for PROSPECTOR – a mixture of Bayesian inference techniques and fuzzy logic. Written in PROLOG and POP-2 in a highly modular fashion. (*Merry, 1983*).

CSRL (Conceptual Structures Representation Language): A language for building hierarchical diagnostic expert systems. Based on the paradigm of "cooperating diagnostic specialists". (*Bylander, Mittal, and Chandrasekaran, 1985*).

DIGS: A domain-independent expert system for the diagnosis of complex electronics. Uses a graph-like model of a diagnostic net and performs breadth-first and depth-first search based on the current symptoms. Developed in PROLOG running under the control of USCD PASCAL. (*Agre and colleagues, 1985*).

EXPERT: (with SEEK: (System for Experimentation with Expert Knowledge)) A FORTRAN based general purpose system that uses findings, hypotheses, and rules relating the findings and hypotheses to obtain a conclusion that is assigned a level of certainty. Developed at Rutgers University.

IDM: (Integrated Diagnostic Model): Integrates "deep" and "shallow" sources of knowledge. Uses a semantic network for shallow knowledge representation and a hierarchical, functional model for deep knowledge. (*Fink, Lusth, and Duran, 1984; Fink and Lusth, 1987*).

PDS (Process Diagnostic System) Application to Turbine-Generators. PDS is an expert shell system developed by M.S. Fox at the Carnegie-Mellon University Robotics Institute. PDS uses a rule base similar to the inference net in PROSPECTOR and has many features similar to MYCIN and PROSPECTOR. Uses certainty factors. Appropriate for on-line, sensor-based diagnosis. (*Gonzalez and Lowenfeld, 1986*).

ROGET: A knowledge-based system for acquiring the structure of a DES. Uses concepts common to most DESs. Developed at Stanford. (*Bennett, 1985*).

7. FUTURE DES RESEARCH AREAS

The areas of future research and development efforts include:

1. DES systems that have learning capabilities, i.e., where the system can improve its performance from experience (see Rowe and collegues {85}; Pazzani, {72}; Hall, Lathrop and Kirk, {40}).

2. Integrating DES into the design process (see Robach, Malecha, and Michel, {84}).

3. Improved uncertainty mechanisms.

4. Multiple fault handling capabilities (see de Kleer and Williams, {25-26}; Edgar and Petty, {31}; Williams, Orgren, and Smith, {105}; Josephson and colleagues, {49}).

5. Language for describing systems for diagnostic purposes (see Bylander, Mittal, and Chandrasekaran, {16}).

6. Improved BIT DES.

7. Adaptable to the current user and/or problem situation (White and Sykes, {103}).

8. Implementation of diagnostic intelligent systems in parallel computing technologies (Narayanan and Viswanadham, {64}).

REFERENCES

1. J.A. Adams, M. Gale, J.W. Dempsey, G.W. Kaizar, N. Straguzzi, "Artificial Intelligence Applications", *RCA Engineer*, 68-75 (Jan.-Feb. 1986).

2. K. Adlassnig, "Fuzzy Set Theory in Medical Diagnosis", *IEEE Transactions on Systems, Man, and Cybernetics*, Vol. SMC-16, No 2, 260-265 (1986).

3. G. Agre, V. Sgurev, D. Dochev, C. Dichev, and Z. Markon, "An Implementation of the Expert System DIGS for Diagnostics", *Computers and Artificial Intelligence*, Vol. 4, No 6, 495-502 (1985).

4. J.S. Aikins, J.C. Kunz, and E.H. Shortliffe, "PUFF: An Expert System for Interpretation of Pulmonary Function Data", *Computers in Biomedical Research*, Vol. 16, 199-208 (1983).

5. J.F. Baldwin, "Fuzzy Sets and Expert Systems", *Information Sciences*, Vol. 36, 123-156 (1985).

6. W. Bandler, and L. Kohout, "Probabilistic Versus Fuzzy Production Rules in Expert Systems", *Int. J. Man-Machine Studies*, Vol. 22, 347-353 (1985).

7. S.W. Barth, and S.W. Norton, "Knowledge Engineering Within a Generalized Bayesian Framework", *Uncertainty in Artificial Intelligence*, AAAI-86 Workshop, Philadelphia, PA, 7-16 (1986).

8. M. J. Bending, "Hitest: A Knowledge-Based Test Generation System", *IEEE Design and Test*, 83-92 (May, 1984).

9. J.S. Bennett, and C.R. Hollander, "DART: An Expert System for Computer Fault Diagnosis", *IJCAI-81*, 843-845 (1981).

10. J.S. Bennett, "ROGET: A Knowledge-Based System for Acquiring the Conceptual Structure of a Diagnostic Expert System", *Journal of Automated Reasoning*, Vol. 1, 49-74 (1985).

11. G. Biswas, and T.S. Anand, "Using the Dempster-Shafer Scheme in a Diagnostic Expert System Shell", *Uncertainty in Artificial Intelligence*, Workshop, Seattle, Washington, 98-107 (July, 1987).

12. L. Bolc, A. Kowalski, M. Kozlowska and T. Strzalkowski, "A Natural Language Information Retrieval System with Extensions Towards Fuzzy Reasoning", *International Journal Man-Machine Studies*, Vol. 23, 335-367 (1985).

13. P.P. Bonissone, and H.E. Johnson, "Expert System for Diesel Electric Locomotive Repair", *IJCAI-83*, (1983).

14. L. Brownston, R. Farrel, E. Kant, and N. Martin, *Programming Expert Systems in OPS5: An Introduction to Rule-Based Programming*, Addison-Wesley, New York (1985).

15. B. Buchanan, and E. Shortliffe, *Rule-Based Expert Systems: The MYCIN Experiments of the Stanford Heuristic Programming Project*, Addison-Wesley, New York (1984).

16. T. Bylander, S. Mittal, and B, Chandrasekaran, "CSRL: A Language for Expert Systems for Diagnosis", *Comp. & Maths. with Appls.*, Vol. 11, No 5, 449-456 (1985).

17. R.R. Cantone, F.J. Pipitone, W.B. Lander, and M.P. Varrone, "Model-Based Probabilistic Reasoning for Electronics Troubleshooting", *IJCAI-83*, 207-211 (1983).

18. B. Chandrasekaran, and S. Mittal, "Deep versus Compiled Knowledge Approaches to Diagnostic Problem-Solving", *Developments in Expert Systems*, Academic Press, London, 23-34 (1984).

19. E. Charniak, and D. McDermott, *Introduction to Artificial Intelligence*, Addison-Wesley, New York (1985).

20. M. Chiu, and E. Niedermeyer, "Knowledge-Based Diagnosis of Manufacturing Cells", *Siemens Research and Development Reports*, Vol. 14, No 5, 230-237 (1985).

21. W.J. Clancey, and E.H. Shortliffe, eds., *Readings in Medical Artificial Intelligence: The First Decade*, Addison-Wesley, Reading, Mass., (1984).

22. R. Davis, "Diagnostic Reasoning Based on Structure and Behavior", *Artificial Intelligence*, Vol. 24, 347-410 (1984a).

23. R. Davis, "Reasoning from First Principles in Electronic Troubleshooting", *Development in Expert Systems*, Academic Press, London, 1-21 (1984b).

24. K. Davis, "DORIS: Diagnostic Oriented Rockwell Intelligent System", *IEEE AES Magazine*, 18-21 (July, 1986).

25. J. de Kleer, and B. Williams, "Reasoning About Multiple Faults", *AAAI-86: Proceedings of the Fifth National Conference on Artificial Intelligence*, Philadelphia, PA, 132-139 (August, 1986).

26. J. de Kleer, and B. Williams, "Diagnosing Multiple Faults", *Artificial Intelligence*, Vol. 32, No 1, 97-130 (April, 1987).

27. A.P. Dempster, "Generalization of Bayesian Inference", *J. Royal Statistical Society*, Series B, Vol. 30, 205-247 (1968).

28. K.L. Downing, "Diagnostic Improvement through Qualitative Sensitivity Analysis and Aggregation", *Proceedings of the Sixth National Conference on Artificial Intelligence: Volume II*, Seattle, Washington, 789-793 (July, 1987).

29. R. Duda, J. Gaschnig, and P. Hart, "Model Design in the PROSPECTOR Consultant System for Mineral Exploration", *Expert Systems in the Microelectronic Age*, Michie. D., (Ed.), University of Edinburg Press, Scotland, 153-167 (1979).

30. A. Dutta, "Reasoning with Imprecise Knowledge in Expert Systems", *Information Sciences*, Vol. 37, 3-24 (1985).

31. G. Edgar, and M. Petty, Location of Multiple Faults by Diagnostic Expert Systems", *Proc. SPIE Int. Soc. Opt. Eng.*, Vol. 485, 39-45 (1984).

32. P.K. Fink, J.C. Lusth, and J.W. Duran, "A General Expert System Design for Diagnostic Problem Solving", *IEEE Proc. Workshop on Principles of Knowledge-Based Systems*, IEEE Comp. Soc., 45-52 (Dec., 1984).

33. P.K. Fink, and J.C. Lusth, "Expert Systems and Diagnostic Expertise in the Mechanical and Electrical Domains", *IEEE Transactions on Systems, Man, and Cybernetics*, Vol. SMC-17, No 3, 340-349 (1987).

34. R. Gabriel, "Lisp Expert Systems are More Useful", *Electronics*, 65 (August, 1986).

35. I. Gazdik, "Fault Diagnosis and Prevention by Fuzzy Sets", *IEEE Transactions on Reliability*, Vol., R-34, No 4, 382-388 (Oct., 1985).

36. M.R. Genesereth, "The Use of Design Descriptions in Automated Diagnosis", *Artificial Intelligence*, Vol. 24, 411-436 (1984).

37. I. Goodman, and H. Nguyen, *Uncertainty Models for Knowledge-Based Systems*, North-Holland, Amsterdam, (1985).

38. A.J. Gonzalez, and S. Lowenfeld, "On-Line Diagnosis of Turbine Generators Using Artificial Intelligence", *IEEE Transactions on Energy Conversion*, Vol. EC-1, No 2, 68-74 (June, 1986).

39. O. Grillmeyer, and A.J. Wilkinson, "The Design and Construction of a Rule Base and an Inference Engine for Test System Diagnosis", *International Test Conference*, 857-867 (1985).

40. R.J. Hall, R.H. Lathrop, and R.S. Kirk, "A Multiple Representation Approach to the Understanding of Time Behavior of Digital Circuits", *Proceedings of the Sixth National Conference on Artificial Intelligence*, Seattle, Washington, 799-803 (July, 1987).

41. R.T. Hartley, "CRIB: Computer Fault-finding Through Knowledge Engineering", *Computer*, 76-83 (March, 1984).

42. F. Hayes-Roth, D. Waterman, and D. Lenat, (eds.), *Building Expert Systems*, Addison-Wesley, New York (1983).

43. F. Hayes-Roth, "The Knowledge-Based Expert System: A Tutorial", *Computer*, 11-28 (Sept., 1984).

44. D. Heckerman, "An Axiomatic Framework for Belief Updates", *Uncertainty in Artificial Intelligence*, Workshop AAAI 86, Philadephia, Pa., 123-128 (1986).

45. M.R. Herbert, and G.H. Williams, "An Initial Evaluation of the Detection and Diagnosis of Power Plant Faults Using a Deep Knowledge Representation of Physical Behavior", *Expert Systems*, Vol. 4, No 2, 90-105 (1987).

46. M. Hofmann, J. Caviedes, J. Bourne, G. Beale, and A. Brodersen, "Building Expert Systems for Repair Domains", *Expert Systems*, Vol. 3, No 1, 4-12 (January, 1986).

47. A.H. Jackson, "Expert Systems for Fault Diagnosis", *MILCOMP '85: Military Computers, Software, and Graphics*, London, England, 449-455 (Oct., 1985).

48. A. Jacobson, "Lisp is Not Needed for Expert Systems", *Electronics*, 64 (August, 1986).

49. J.R. Josephson, B. Chandrasekaran, J.R. Smith, and M.C. Tanner, "A Mechanism for Forming Composite Explanatory Hypotheses", *IEEE Transactions on Systems, Man, and Cybernetics*, Vol. SMC-17, No 3, 445-454 (1987).

50. J. Kacprzyk, and R.R. Yager, "Emergency-Oriented Expert Systems: A Fuzzy Approach", *Information Sciences*, Vol. 37, 143-155 (1985).

51. C. Kalme, "Decision Under Uncertainty in Diagnosis", *Uncertainty in Artificial Intelligence*, Workshop AAAI 86, Philadelphia, 145-1502 (1986).

52. L. Kanal, and J. Lemmer, (eds.), *Uncertainty in Artificial Intelligence*, North-Holland, Amsterdam (1986).

53. J.J. King, "Artificial Intelligence Techniques for Device Troubleshooting", Computer Science Laboratory Technical Note Series, Hewlett-Packard Co., Palo Alto, (August, 1982).

54. J. Kitowski, and I. Ksiazek, "Fuzzy Logic Applications for Failure Analysis and Diagnosis of a Primary Circuit of the HTR Nuclear Power Plant", *Computer Physics Communications*, Vol. 38, 323-327 (1985).

55. T.J. Laffey, W.A. Perkins, and T.A. Nguyen, "Reasoning About Fault Diagnosis with LES", *First Conference on Artificial Intelligence Applications*, 267-273 (1984).

56. Liu, Chen-Ching, and Tomsovic, Kevin, "An Expert System Assisting Decision-Making of Reactive Power/Voltage Control", *IEEE Transactions on Power Systems*, Vol. Pwrs-1, No 3, 195-201 (August, 1986).

57. F.E. Masarie, R.A. Miller, and J.D. Myers, "INTERNIST-1 Properties: Representing Common Sense and Good Medical Practice in a Computerized Medical Knowledge Base", *Computers and Biomedical Research*, Vol. 18, 458-479 (1985).

58. D. McDermott, and R. Brooks, "ARBY: Diagnosis with Shallow Causal Models", *AAAI-82*, 370-372 (1982).

59. Martin, Merry, "APEX 3: An Expert System Shell for Fault Diagnosis", *The GEC Journal of Research*, Vol. 1, No 1, 39-47 (1983).

60. R. Milne, "Strategies for Diagnosis", *IEEE Transactions on Systems, Man, and Cybernetics*, Vol. 17, No 3, 333-339 (1987).

61. R.A. Miller, H.E. Pople, and G.D. Myers, "INTERNIST-1: An Experimental Computer-Based Diagnostic Consultant for General Internal Medicine", *New England Journal of Medicine*, Vol. 307, 468-476 (1982).

62. G. Montgomery, "Artificial Intelligence Applied to Aircratf Integrated Diagnostics", Flight Dynamics Laboratory, Air Force Wright Aeronautical Laboratories, Wright-Patterson AFB, (May, 1986).

63. R.E. Moon, S.Z. Jordanov, I.B. Turksen, and A. Perez, "Human-Like Reasoning Capability in a Medical Diagnostic System: The Application of Fuzzy Set Theory to Computerized Diagnosis", *Journal of Clinical Computing*, Vol. 8, No 3, 122-151 (1979).

64. N.H. Narayanan, and N. Viswanadham, "A Methodology for Knowledge Acquisition and Reasoning in Failure Analysis of Systems", *IEEE Transactions on Systems, Man, and Cybernatics*, Vol. SMC-17, No 2, 274-288 (1987).

65. C.V. Negotia, *Expert Systems and Fuzzy Systems*, Benjamin/Cummings, (1985).

66. W.R. Nelson, "REACTOR: An Expert System for Diagnosis and Treatment of Nuclear Power Accidents", *AAAI: Proceedings of the National Conference on Artificial Intelligence*, Pittsburgh, PA, (1982).

67. S. Norton, "An Explanation Mechanism for Bayesian Inferencing Systems", *Uncertainty in Artificial Intelligence*, Workshop AAAI 86, Philadelphia, Pa., 145-1502 (1986).

68. P. Odryna, and A. Strojwas, "PROD: A VLSI Fault Diagnosis System", *IEEE Design & Test*, 27-35 (December, 1985).

69. B.L. Palowitch, and M.A. Kramer, "The Application of a Knowledge-Based Expert System to Chemical Plant Fault Diagnosis", *Proceedings on the 1985 American Control Conference*, 646-651 (June, 1985).

70. J.Y. Pan, "Qualitative Reasoning with Deep-Level Mechanism Models for Diagnoses of Mechanism Failures", *First Conference on Artificial Intelligence Applications*, 295-301 (1984).

71. L.F. Pau, "Survey of Expert Systems for Fault Detection, Test Generation and Maintenance", *Expert Systems*, Vol. 3, No 2, 100-111 (April, 1986).

72. M. Pazzani, "Refining the Knowledge Base of a Diagnostic Expert System: An Application of Failure Driven Learning" *AAAI-86: Proceedings*

of the Fifth National Conference on Artificial Intelligence, Philadelphia, PA, 1029-1035 (1986).

73. M. Pazzani, "Failure Driven Learning of Fault Diagnostic Heuristics", *IEEE Transactions on Systems, Man, and Cybernetics*, Vol. SMC-17, No 3, 380-394 (1987).

74. J. Pearl, "On the Logic of Probabilistic Dependencies", *AAAI-86: Proceedings of the Fifth National Conference on Artificial Intelligence*, Philadelphia, PA, 339-343 (1986).

75. Y. Peng, and J.A. Reggia, "A Probabilistic Causal Model for Diagnostic Problem Solving, Part I: Integrating Symbolic Causal Inference with Numeric Probability Inference", *IEEE Transactions on Systems, Man, and Cybernetics*, Vol. SMC-17, No 2, 146-162 (1987).

76. Y. Peng, and J.A. Reggia, "A Probabilistic Causal Model for Diagnostic Problem Solving, Part II: Diagnostic Strategy", *IEEE Transactions on Systems, Man, and Cybernetics*, Vol. SMC-17, No 3, 395-406 (1987).

77. F. Pipitone, "The FIS Electronics Troubleshooting System", *Computer*, 68-76 (July, 1986).

78. C. Powell, C. Pickering, and P. Westcourt, "System Integration of Knowledge-Based Maintenance Aids", *AAAI-86: Proceedings of the Fifth National Conference on Artificial Intelligence*, Philadelphia, PA, 851-855 (1986).

79. A.L. Ralescu, and D.A. Ralescu, "Probability and Fuzziness", *Information Sciences*, Vol. 34, 85-92 (1984).

80. R. Reiter, "A Theory of Diagnosis from First Principles", *Artificial Intelligence*, Vol. 32, No 1, 57-96 (1987).

81. J.A. Reggia, D.S. Nau, and P.Y. Wang, "Diagnostic Expert Systems Based on a Set Covering Model", *Developments in Expert Systems*, Academic Press, London, 35-58 (1984).

82. J.A. Reggia, D.S. Nau, and P.Y. Wang, "A Formal Model of Diagnostic Inference. I. Problem Formulation and Decomposition", *Information Sciences*, Vol. 37, 227-256 (1985).

83. J.A. Reggia, D.S. Nau, P.Y. Wang, and Y. Peng, "A Formal Model of Diagnostic Inference. II. Algorithmic Solution and Applications", *Information Sciences*, Vol. 37, 257-285 (1985).

84. C. Robach, P. Malecha, and G. Michel, "GATA: A Computer-Aided Test Analysis System", *IEEE Design and Test*, 68-79 (May, 1984).

85. M. Rowe, A. Veitch, R. Keener, and B. Lantz, "An Adaptive/Learning Diagnostic System for Complex Domains: ET", *TI Engineering Journal*, Vol. 3, No 2, 67-72 (1986).

86. P.M. Ryan, and A.J. Wilkinson, "Knowledge Acquisition for ATE Diagnosis", *IEEE AES Magazine*, 5-12 (1986).

87. E.A. Scarl, J.R. Jamieson, and C.I. Delaune, "Diagnosis and Sensor Validation Through Knowledge of Structure and Function", *IEEE Transactions on Systems, Man, and Cybernetics*, Vol. SMC-17, No 3, 360-368 (1987).

88. S. Schwartz, J. Baron, and J. Clark, "A Causal Bayesian Model for the Diagnosis of Appendicitis", *Uncertainty in Artificial Intelligence*, Workshop AAAI 86, Philadelphia, PA, 229-236 (1986).

89. R.D. Shachter, and L.J. Bertrand, "Efficient Inference on Generalized Fault Diagrams", *Uncertainty in Artificial Intelligence*, Workshop, Seattle, Washington, 413-420 (July, 1987).

90. G. Shafer, *A Mathematical Theory of Evidence*, Princeton University Press, Princeton, NJ, (1976).

91. E.H. Shortliffe, B.G. Buchanan, and E.A. Feigenbaum, "Knowledge Engineering for Medical Decision Making: A Review of Computer-Based Clinical Decision Aids", *Proceedings of the IEEE 67*, 1207-1224 (1979).

92. H. Shubin, and J.W. Ulich, "IDT: An Intelligent Diagnostic Tool", *Proc. AAAI-82*, 290-295 (1982).

93. P. Snow, "Bayesian Inference Without Point Estimates", *AAAI-86: Proceedings of the Fifth National Conference on Artificial Intelligence*, Philadelphia, PA, 233-237 (1986).

94. J. Sticklen, B. Chandrasekaran, J.W. Smith, and J. Suirbely, "A Comparison of the Diagnostic Subsystems of MDX and MYCIN", *IEEE Proc. Workshop on Principles of Knowledge-Based Systems*, IEEE Comp. Soc., 205-212 (Dec., 1984).

95. S.N. Talukdar, E. Cardazo, and L.V. Leao, "Toast: The Power System Operator's Assistant", *Computer*, 53-60 (July, 1986).

96. S.N. Talukdar, E. Cardozo, and T. Perry, "The Operator's Assistant-An Intelligent, Expandable Program for Power System Trouble Analysis", *IEEE Transactions on Power Systems*, Vol. PWRS-1, No 3, 182-187 (Aug., 1986).

97. M. Thandasseri, "Expert Systems Application for TXE4A Exchanges", *Electrical Communication*, Vol. 60, No 2, 154-161 (1986).

98. T.F. Thompson, and R.M. Wojcik, "MELD: An Implementation of a Meta-Level Architecture for Process Diagnosis", *First Conference on Artificial Intelligence Applications*, 321-330 (1984).

99. T.B. Turksen, "Fuzzy Reasoning in Medical Diagnosis", *Modelling and Simulation - Proceedings of the Annual Pittsburgh Conf.*, Vol. 14, 1029-1033 (1983).

100. S. Umeyama, "The Complementary Process of Fuzzy Medical Diagnosis and Its Properties", *INFORMATION SCIENCES 38*, 229-242 (1986).

101. D.L. Waterman, *A Guide to Expert Systems*, Addison-Wesley, New York, (1986).

102. C.C. White, "A Posteriori Representations Based on Linear Inequality Descriptions of a Priori and Conditional Probabilities", *Systems, Man, and Cybernetics*, Vol. SMC-16, No 4, 570-572 (1986).

103. C.C. White, and E.A. Sykes, "A User Preference Guided Approach to Conflict Resolution in Rule-Based Systems", *IEEE Trans. Systems, Man, and Cybernetics*, Vol. SMC-16, No 2, 276-278 (1986).

104. A.J. Wilkinson, "MIND: An Inside Look at an Expert System for Electronic Diagnosis", *IEEE Design and Test*, 69-77 (Aug., 1985).

105. T.L. Williams, P.J. Orgren, and C.L. Smith, "Diagnosis of Multiple Faults in a Nationwide Communications Network", *Eighth International Joint Conference on Artificial Intelligence*, Vol. 1, 179-181 (1983).

106. R. Yadrick, D. Vaughan, B. Perrin, P. Holden, and K. Kempf, "Evaluation of Uncertain Inference Models I: PROSPECTOR", *Unceratinty in Artificial Intelligence*, Workshop AAAI 86, Philadelphia, PA, 333-338 (1986).

107. J. Yen, "A Reasoning Model Based on an Extended Dempster-Shafer Theory", *AAAI-86: Proceedings of the Fifth National Conference on Artificial Intelligence*, Philadelphia, PA, 125-131 (1986).

108. L. Zadeh, "Fuzzy Sets", *Inform. and Control*, Vol. 8, 338-353 (1965).

109. L. Zadeh, and A. Ralescu, "On the Combinality of Evidence in the Dempster-Shafer Theory", *Uncertainty in Artificial Intelligence*, Workshop AAAI 86, Philadelphia, PA, 347-349 (1986).

110. G. Zwingelstein, and B. Monnier, "Artificial Intelligence Applications to the Surveillance and Diagnostics of Nuclear Power Plants", *Trans. Amer. Nucl. Soc.*, Vol. 50, 515-516 (1985).

CONTRIBUTORS

S. ABU EL ATA-DOSS   ADERSA, Verrieres Le Buisson, BP 52, Cedex, France.

G.O. BEALE   Department of Electrical and Computer Engineering, George Mason University, Fairfax, Virginia, USA.

G. BENGTSON   Applied Electronics, Chalmers University, Gothenburg, Sweden.

P. BERTOK   Computer and Automation Institute, Hungarian Academy of Sciences, Budapest, Hungary.

J. BOURNE   The Center for Intelligent Systems, Vanderbilt University, Box 1570, Station B, Nashville, TN, USA.

A. BRODERSEN   The Center for Intelligent Systems, Vanderbilt University, Box 1570, Station B, Nashville, TN, USA..

C. CARTWRIGHT   Cambridge Consultants Limited, England.

J. CAVIEDES   Philips Laboratories, Briarcliff, Manor, New York, USA.

F.E. FINCH   Department of Chemical Engineering, MIT, Cambridge, MA 02139, USA.

I. GAZDIK   Elinsborgsbacken 23, S-163 64 Spanga, Sweden.

N. GIAMBIASI   L.E.R.I., Parc d'Activités Scientifiques et Techniques, 3000 Nimes, France.

J.-D. KATZBERG   Electronic Information Systems Engineering, University of Regina, Regina, Saskatchewan, Canada.

K. KAWAMURA   Center for Intelligent Systems, Vanderbilt University, Box 1570, Station B, Nashville, Tennessee, USA.

M.A. KRAMER   Department of Chemical Engineering, MIT, Cambridge, MA 02139, USA.

T. LI   Department of Computer Science, The University of Adelaide, G.P.O. Box 498, Adelaide, Australia.

J. MAROLD   Centre Recherche en Informatique de Nancy (CRIN), Vandoeuvre-lès-Nancy, France.

P. OSBORNE   The Center for Intelligent Systems, Vanderbilt University, Box 1570, Station B, Nashville, TN, USA.

C. OUSSALAH   L.E.R.I., Parc d'Activités Scientifiques et
                       Techniques, 3000 Nimes, France.

G. PAPAKONSTANTINOU   Computer Science Division, National Technical
                       University of Athens, Zografou 15773, Athens,
                       Greece.

A. PATERSON   PA Computers and Telecommunications, Rochester
                       House, London.

L.F. PAU   Technical University of Denmark, DK 2800 Lyngby,
                       Denmark.

D. REYNOLDS   Cambridge Consultants Limited, England.

A. ROSS   The Center for Intelligent Systems, Vanderbilt
                       University, Box 1570, Station B, Nashville, TN, USA.

P. ROUX   CIMSA-SINTRA, 10 Avenue de l'Europe, 78140 Vélizy,
                       France.

P. SACHS   PA Computers and Telecommunications, Rochester
                       House, London.

J.F. SANTUCCI   L.E.R.I., Parc d'Activités Scientifiques et
                       Techniques, 3000 Nimes, France.

J.D. SCHAFFER   The Center for Intelligent Systems, Vanderbilt
                       University, Box 1570, Station B, Nashville, TN, USA.

W.T. SCHERER   Department of Systems Engineering, University of
                       Virginia, Thornton Hall, Charlottesville, VA, USA.

S.N. SRIHARI   Department of Computer Science, State University
                       of New York at Buffalo, Buffalo, New York, USA.

S. TZAFESTAS   Computer Science Division, National Technical
                       University of Athens, Zografou 15773, Athens,
                       Greece.

J. VANCZA   Computer and Automation Institute, Hungarian
                       Academy of Sciences, Budapest, Hungary.

C.C. WHITE, III   Department of Systems Engineering, University of
                       Virginia, Thornton Hall, Charlottesville, VA, USA.

W. ZIARKO   Department of Computer Science, University of
                       Regina, Regina, Saskatchewan, Canada.

INDEX